W9-CPF-907

ASSESSMENT IN TECHNICAL
AND
PROFESSIONAL COMMUNICATION

Edited by

Margaret N. Hundleby
University of Toronto

and

Jo Allen
Widener University

Baywood's Technical Communications Series
Series Editor: CHARLES H. SIDES

Baywood Publishing Company, Inc.
AMITYVILLE, NEW YORK

Copyright © 2010 by Baywood Publishing Company, Inc., Amityville, New York

Baywood Publishing Company, Inc.

26 Austin Avenue
P.O. Box 337
Amityville, NY 11701
(800) 638-7819
E-mail: baywood@baywood.com
Web site: baywood.com

Library of Congress Catalog Number: 2009019357
ISBN 978-0-89503-379-6

Library of Congress Cataloging-in-Publication Data

Assessment in technical and professional communication / edited by Margaret N. Hundleby and Jo Allen.
      p. cm. -- (Baywood's technical communications series)
   Includes bibliographical references and index.
   ISBN 978-0-89503-379-6 (cloth : alk. paper) 1. Communication of technical information.
2. Technical writing--Study and teaching. I. Hundleby, Margaret N., 1940- II. Allen, Jo.
   T10.5.A84 2009
      601'. 4--dc22

                                                    2009019357

Table of Contents

ASSESSING INTERCULTURAL/
INTERNATIONAL PROJECTS

The Assessment Landscape in Technical and Professional Communication: Evolutionary Thinking and Practice in an Emerging Imperative

Margaret Hundleby, University of Toronto
Jo Allen, Widener University

The purpose of this book is straightforward: to address the state of assessment of learning in Technical and Professional Communication. On the one hand, writing has played such a crucial, universal role in assessing all sorts of student learning outcomes that we bear some legitimacy from a sense of history and longevity. We know about assessment, in other words, because our students' work has been such an integral part of both the documentation and the demonstration of learning—both the intellectual and physical evidence, as it were. On the other hand, however, assessment in our field has suffered both from irregular attention to its status in our overall practice and from uncertainty about productive and authentic strategies. Like our near kin in composition programs, we still face a number of unanswered questions about *why* we do as we do. An even greater number of questions exist about *how* we assess the work of our students to foster the success of our programs.

These conditions stem partly from the customary status of assessment as the stepchild of pedagogical activity, tended to after the more important members of the instructional family—curriculum design, instructional technology applications, course delivery—have been nourished. A second difficulty arises in the somewhat orphaned condition of technical and professional communication; standing in line behind mainstream composition, it is often seen as something of an add-on in English studies and, therefore, without even the benefit of concern for maintenance and improvement of a core requirement.

Despite the distinctive progress made toward establishing Technical Communication as a recognizable, stable, and highly able field—especially in recent years—we have not developed the fully functioning assessment practices that would mark our ability to explain our discipline and our practices to the rest of the academic world or commercial enterprise, or to demonstrate our efficient and effective use of resources in a knowledge-making society. Not least because these moves ensure access to a highly desirable, even necessary, legitimizing of status in and out of the academy, we need urgently to begin the process of shifting the status of our assessment practices to a level commensurate with the effort we are putting into building the status of the field, which arguably will be diminutive until we address the assessment component. As such, it must become more "front-loaded" into our thinking and planning, even driving the other processes that build the curriculum and, thus, the discipline.

GETTING TO MEANINGFUL ASSESSMENT

Very little material is available to help us understand exactly what we need to do to begin this work. As a rule, we tend to borrow general methodologies from Rhetoric and Composition, which have in turn borrowed them from Educational Assessment (Lynne, 2004). Originally an approach depending heavily on measurement theory, writing evaluation historically used a constrained-response format such as multiple choice questions, testing devices that did little more than satisfy the most elemental expectations of measurement theory; especially because the main purpose was that of rating writing abilities efficiently across a population that had to be categorized for global decisions such as placement. Subsequently, composition scholars interested in direct assessment undertook to establish holistic scoring as a replacement methodology, arguing not only that this means establishes a reliable way to rank students, but also that it has a strong relational effect to teaching and learning writing and could serve as a useful means of helping teachers understand how assessment justifies their pedagogical practice (Cooper & Odell, 1977; Godschalk, Swineford, & Coffman, 1966; White, Lutz, & Kamusikiri, 1986).

As Brian Huot (2002) notes, however, any goal-directed approach that attends solely to solving technical problems can backfire by narrowing the scope of the assessment so that it merely lines up with the decisions that will be made by the assessment sponsors—whether they know anything about the substance and character of the discipline under scrutiny or not. We need, Huot suggests, to distance ourselves from the narrowly conceived concerns of standardized assessment that focus on attaining mere technical efficiency. But even more centrally, we need to be aware that any assessment practice is underpinned by theory, or principles, that either already motivates our practices or enables us to create new ones "more consistent with the theories we hold or want to hold" (p. 168). Presumably, also, we will be able to gain the outcomes we want to gain, take back the authority of our own practice, and reconceive the meaningfulness of assessment on the basis of what we value as necessary and functional to literacy.

While the story of the development of writing assessment current in mainstream composition is inevitably one of twists, turns, and byways, it also shows a remarkably steady movement along a path that is of considerable interest to Technical Communication—the characterization of writing instruction as dependent on the social construction of knowledge. In company with Huot, most theorists working on writing or assessment or both during the late 1980s and across the 1990s have incorporated social constructivism into their analysis and discussion as the key element in what is concluded to be the authentic, productive, and coordinated practice of teaching and assessing writing (Huot, 2002; Lynne, 2004). This viewpoint predicates two fundamental concerns: one is understanding the existence of multiple literacies—"the diversity of the reading and writing abilities necessary in various circumstances . . . [along with] the notion that different situations exhibit different values" (Lynne, 2004, p. 57)—which requires that any assessment effort places both the assessment and the work assessed into its context of operation; the other is acknowledging the substance of the assessment, combining "the content or subject matter of an assessment . . . [with] the reasons for and the object(s) of assessment" (p. 122)—a task that similarly requires taking account of the disciplinary expectations for what counts as both knowledge and literacy within its terms of definition.

Taking these concerns into account makes it crucial to ground any approach in an appropriate disciplinary locus. In other words, meaningful practice is possible if, and only if, an established body of knowledge guides identification, analysis, and interpretation of the information generated in the process of assessment. Doing so permits us "to claim the authority to define the principles by which to describe, evaluate and re-imagine what evidence of literate ability— as well as assessment itself—looks like" (Lynne, 2004, p. 168); but it also carries us well beyond routines of agreement on what criteria to use to evaluate an assignment, curriculum, or program. Deeply reflective practice yields three consequences: the first is the necessary abolition of the notion that a set, and static, methodology can be invoked at all times; the second is that renovating expectations for assessment provides a starting place for Technical Communication to think about the overall state of its pedagogy; the third is that questioning what is germane to the epistemic conditions and knowledge consequences informs the specifics of both our instruction and evaluation practices. Getting to meaningful assessment results in new power and authenticity for Technical Communication—and in an equal burden of responsibility for establishing and maintaining its practice.

ACCOUNTING FOR EPISTEMIC VALUES

Inherent in the renovated view of assessment we are considering here is the responsibility we all now share for ensuring that our pedagogical practices are based on an understanding of the epistemic values that characterize our discipline. If we don't know what has value in our practice, we can't teach it adequately or assess for it authentically. From the early years of asking "What's Technical about Technical Writing?" (Dobrin, 1983), we have necessarily been concerned with exactly what it means to communicate for technical and professional purposes. One of the most fundamental reasons for this concern is simply that a major disciplinary expectation is being able to understand and apply new strategies—whether graphic, verbal, or media-related. This situation places us well ahead of our composition colleagues in seeing the need for a thorough understanding of multiple literacies and the resulting responsibility to operate within the frameworks of their distinctive discourses. Similarly, we have participated in the reality of being required to acknowledge and work with the context-specific character of communication undertakings. From the early days of discourse and genre theory situated vis-à-vis communication practices for technical and scientific purposes (Anderson, Brockmann, & Miller, 1983; Bazerman, 1988) to later studies of rhetorical and cultural concerns within the field (Bolter, 2001; Star, 1995), we have understood how context affects both the configuration of our epistemology and the epistemic character of our practices. Thus, both the concept of multiple literacies and the comprehension of the significance of contextualization figure as central principles in constructing the authority necessary to reclaim for ourselves the practice of meaningful assessment.

A second factor in determining epistemic values in our field as we aim for assessment that is both authoritative and meaningful is the effect on the characterization of Technical Communication's substance that derives from the Social Study of Science and from the correlated work of Activity Theory. Scholarship in these areas investigates both the motivation and consequences of doing the work of science and technology. This understanding of practice contributes to our coming to terms with the purposefully constructed character of the work we write about, as Dorothy Winsor (1987) delineates in *Writing Like an Engineer*. Even more importantly, these approaches take seriously the contribution of Technical Communication to the accomplishment of the work: through visual as well as verbal formats, we represent how the work takes place and thus increase the opportunities for identifying what the actual processes are, exactly why they proceed as they do, and whether or not the outcomes support any claims to significance (Chaiklin & Lave, 1996; Latour, 1987; Lynch & Woolgar, 1991). Undertaking to write, speak, or otherwise represent work as purposeful activity in support and furtherance of desired outcomes—development and discovery in science settings, say, or problem solving in engineering and technology—forms for us an equally purposeful activity of making the knowledge available and readily comprehensible. The most telling part of this arrangement is that the relation is reciprocal: first, an exchange of benefits occurs between science/ technology and communication as they share mutual interests in substantive knowledge-making. Additionally, there is the influence-building between substance and context resulting from developing representation of any instance of work and examining the implications of its constituent activities. As a result, we gain an increased sense of the purposefulness of undertaking the task of communicating about science and technology, but we also acquire a new responsibility for developing the standards and values crucial to conducting the meaningful assessment that will support Technical Communication's ongoing development.

CONDUCTING A NATURALISTIC INQUIRY

In order to come to terms with the gaps we see in assessment undertakings for the field, we have chosen to adapt the advice for educational evaluation presented by Guba and Lincoln in their development of what they call "responsive" or "fourth generation" evaluation (1981, 1989). This work supports and extends the general viewpoint on developing meaningful evaluation that both Huot and Lynne advocate, but also most importantly provides a means of locating the principles and goals within the locus of participation. Guba and Lincoln argue for a two-part process: first, taking account of the needs and expectations of the participants—both evaluators and evaluands—and second, designing the procedures to ensure usefulness of the outcomes in local as well as global terms. Calling the approach "naturalistic inquiry" (1981), they advise building an inventory of criteria that fit the situation and goals of assessment and avoiding predetermined, fixed processes to concentrate on what the outcomes and significance of the outcomes are.

The orientation found in Guba and Lincoln's work fits neatly with Patricia Lynne's (2004) observation that writing assessment needs to be aware of the central role played by defining its own standards and values based on our obvious possession of expert knowledge, particularly paying attention to their being "the product of an ongoing community discussion that incorporates the concerns of interested parties" (p. 70). To benefit and integrate both sets of advice, we have undertaken to provide a starting point for discussion of the possibilities of outlining assessment practices with a high degree of "fit" to Technical Communication by setting up a naturalistic inquiry, and by initiating what we would like to offer as a national conversation on the topic. Indeed, this book may be understood as a vehicle that will lead to conversations began for pointed reasons of the sort that we encounter repeatedly on discussion lists such as those for the Association of Teachers of Technical Writing or the Council of Programs for Technical and Professional Communication, and at conventions, during committee meetings, and even in hallways and offices. This motif is no surprise in a field characterized by being public and collaborative, but it is also notable that such a conversation promises to be meaningful and applicable for the circumstances of teaching and practices that are its substance.

The conversation does not aim particularly at theorizing assessment, nor is there a particular focus on creating models—though at least one formal model and a number of suggestions for rubrics are included. Instead, we are seeking to capitalize on the reciprocal relationship between theory and practice in most literacy efforts. In reading the following chapters, we want readers to think about practice as they look toward theory, and to think about the theoretical underpinnings that begin to emerge as they examine the results of day-to-day assessment practice. Besides acknowledging the fundamental relationship between the two elements, the chapters together and separately aim at inviting all teachers/assessors to think about "fit" for the context in which they operate, both local and global. This approach, we believe, underscores the interconnectedness between all possible pairings of concepts and practitioners that formally define our field.

Our selection of contents also reflects a conviction about areas that, frankly, need the most attention at this initial stage. These have been chosen for their resonance with Technical Communication practice as we know it and with an eye to drawing on the experience of a number of our colleagues who have been doing this work over a sustained period of time. Contributions have been requested from people with specific experience in the substantive areas selected. They all have practiced their art in a wide variety of settings and have been willing to share their self-awareness as well as outcomes naturalized in contexts familiar to—if not necessarily, or even probably, a replica for—the ones each of us experiences. The result is that we have been able to assemble a wide-ranging conversation that foregrounds what strategies often work, what outcomes can usually be expected, and what further implications are likely to appear—the most natural possible benefit of naturalistic inquiry.

JOINING THE CONVERSATION

What you can do with this volume will depend on what you need to take up in the context of your own work and goals. In reviewing the conversation—and taking advice, again, from Guba and Lincoln (1981)—we find a number of different kinds of strategies offered: some fall into the "operational" category, where you will find conversations about both formative goals—developing course content, modifying and improving design, and fitting work to the local context; and summative approaches—valuing outcomes, warranting entities, and certifying for local use. Others fall into the "evaluative" camp, including exchanges on determining merit, or global (disciplinary) sufficiency/degree and on determining worth through local (course and program) valuation. All go well beyond the consideration of evaluation and assessment as "audit" undertakings, limited to taking an inventory of what is present. Instead, they exhibit a well-grounded understanding of the continuing need for situated interaction and informed negotiation that "treats social, cultural and political features as the properties of all human circumstance" (Guba & Lincoln, 1989, p. 253) and makes them a central part of the evaluation process.

To maintain the dynamic of conversation while recording the outcomes of the evaluation inquiries featured here, we have emphasized "pairing," though not necessarily actual dialogue. The first section features the most complicated instance of this arrangement, consisting of two chapters each, with a single essay giving an account of both the specific activities and the specific complexities that underlie the daily round of evaluating students. Taken separately, each explores the scene and actions of a major evaluation project. Taken in tandem, the pair provides a multidimensional look at how the ultimate operation and value of assessment depends on the viewpoint(s) established, the actors involved, and the outcomes sought. Seeing the two discussions in proximity, we have the opportunity to build historical awareness of our roles as agents, of the consequences of our actions in using given techniques for evaluation, and of the overriding need to be aware of the purpose of what we are doing, as we do it.

Following this very Burkean (1969) beginning, the subsequent sections are organized as a single chapter each, but as ones containing two accounts that form a reciprocal pair: one initiates the conversation by addressing a topic from a chosen viewpoint, and the other may respond, extend, extrapolate, redirect, or otherwise react to what has been articulated—all in the spirit of conversing on the topic of assessment. In the first of the chapter pairs, for example, Jo Allen and Paul Anderson draw on their individual and collective wealth of experience to reinforce observations on the role of assessment in institutional accountability. Similar questions of assessment as part of both internal and external accountability—Cargile Cook/Zachry and Dubinsky on portfolios and Carter and Youra on, respectively, an accrediting agency and professional governance—highlight the importance and impact of accounting for the context of the assessment and the purpose of its proceedings. Whether the pairs are closely aligned, as in Jablonski/Nagelhout's and Hart-Davidson's presentations of how assessment acts as a "spinal column" for the design of program conduct and content, or diverge as widely as do Coppola/Elliot's account of their empirical study of assessing graduate students and Savage's observations on the intellectual and political implications inherent in a seemingly straightforward undertaking.

The last section examines assessment undertakings not so much by direct exchange as by parallel accounts of the action carried out from different, and distinctive, viewpoints. Thus we have Starke-Meyerring/Andrews and Bosley's documentation of the opportunities and pitfalls of conducting assessment across cultural divides. One occurs between the United States and Canada, where a

North American orientation draws students and faculty together in terms of operation, but also questions the character of expectations for outcomes. The other, set between the United States and France, emphasizes the effect of differences in procedural concerns, while aligning fairly closely on what their consequences will be. And finally, in the Afterword, Dragga explores the often-overlooked ethical aspects of our work, with "[its] fluidity of data, the sifting of information, and the dynamics of practice," as we are challenged to make meaningful judgments about it.

Although the coverage is hardly exhaustive, and participation is designed to be representative rather than iconic, the conversation initiated here will, we believe, make it possible to enter into and continue the serious consideration of assessment in Technical Communication that has needed attention for quite a while. To do so in conversation with friends and colleagues will make it even more likely to open up further possibilities for an improved understanding of the evaluation as a process with distinctive consequences and a heightened appreciation of our own roles as participants in making meaning through that process.

REFERENCES

Anderson, P. V., Brockmann, R. J., & Miller, C. R. (Eds.). (1983). *New essays in technical and scientific communication: Research, theory, practice.* Amityville, NY: Baywood.

Bazerman, C. (1988). *Shaping written knowledge.* Madison, WI: University of Wisconsin Press.

Bolter, J. (2001). *Writing space* (2nd ed.). Mahwah, NJ: Lawrence Erlbaum Associates.

Burke, K. (1969). *A grammar of motives.* Berkeley, CA: University of California Press.

Chaiklin, S., & Lave, J. (1996). *Understanding practice. Perspectives on activity and context.* Cambridge, UK: Cambridge University Press.

Cooper, C., & Odell, L. (Eds.). (1977). *Evaluating writing. The role of teachers' knowledge about text, learning, and culture.* Urbana, IL: National Council of Teachers of English.

Dobrin, D. (1983). What's technical about technical writing? In P. V., Anderson, R. J. Brockmann, & C. R. Miller (Eds.), *New essays in technical and scientific communication: Research, theory, practice* (pp. 227–250). Amityville, NY: Baywood.

Godshalk, F., Swineford, F., & Coffman, W. (1966). *The measurement of writing ability—A significant breakthrough.* New York: College Entrance Examination Board.

Guba, E., & Lincoln, Y. (1981). *Effective evaluation.* Newbury Park, CA: Sage.

Guba, E., & Lincoln, Y. (1989). *Fourth generation evaluation.* Newbury Park, CA: Sage.

Huot, B. (2002). *(Re)articulating writing assessment for teaching and learning.* Logan, UT: Utah State University Press.

Latour, B. (1983). *Science in action.* Cambridge, MA: Harvard University Press.

Lynch, M., & Woolgar, S. (1991). *Representation in scientific practice.* Cambridge, MA: MIT Press.

Lynne, P. (2004). *Coming to terms. Theorizing writing assessment in composition studies.* Logan, UT: Utah State University Press.

Miller, C. (1984). Genre as social action. *Quarterly Journal of Speech, 70,* 151–167.

Star, S. (Ed.). (1996). *Introduction, ecologies of knowledge.* Albany, NY: SUNY Press.

White, E. M., Lutz, W. D., & Kamusikiri, S. (Eds.). ([1986]1996). *Assessment of writing: Politics, policies, practice.* New York: Modern Language Association of America.

Winsor, D. (1987). *Writing like an engineer.* Mahwah, NJ: Lawrence Erlbaum Associates.

Knowing Where
We Are

CHAPTER 1

Assessment in Action:
A Möbius Tale

Chris M. Anson
North Carolina State University

Whether as preteens with a dose of curiosity or as adults that science somehow passed by, most people's first experience with a Möbius strip is briefly but intensely engaging. A plane, a two-dimensional surface, that *reverses itself.* That loops outward, or inward, over, under—but at what point? The finger is always first, tracing its way along the blade of the strip, curving around its racetrack course, whose angled turns dissolve into the experience of the continuum. Escher-like, it defies perspective and resolution, a small marvel of design, a *trompe l'oeil,* somehow grounded in the hard physics of space.

Figuring out how well we are doing in our programs and curricula usually invokes a loop of activity, perhaps best illustrated by the famous cycle that moves from the articulation of learning outcomes to their assessment and then to "gap-closing"—to implementation and change strategies (Nichols, 1995, p. 50). Typically, program leaders define what they want learners to achieve, figure out a way to measure it, and improve their curriculum from the results. Most often, this process is orchestrated from above and assumes the eventual cooperation of instructors, who put in place the necessary pedagogical changes warranted by the assessment.

More powerful kinds of program assessment, however, such as those described in the varied contributions to this collection, are better represented by a Möbius strip than a simple loop or circle. While a program is developing and assessing outcomes for the students who finish its courses or prescribed curriculum, individual teachers in that curriculum, through their assignments, classroom interactions, study of and response to student work, consultations, and tutorials,

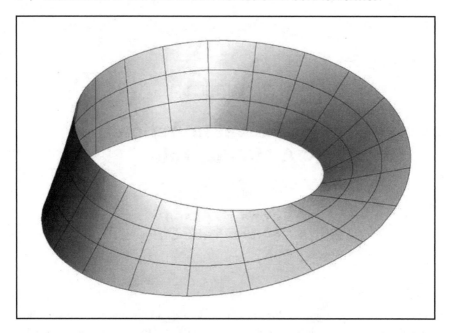

"Möbius strip" visual created in 3D-Xplor Matu.
Used with permission by Dick Palais.

are tracking progress and making decisions about the effectiveness of their teaching. If we think most richly about the processes of assessment, we need to imagine jumping on a Möbius strip and moving from the outside in *and* from the inside out. The two directions end up merging in the same loop, dividing and conjoining at different points; but both are crucial to success. No program can improve from the unarticulated and disparate efforts of a confederation of teachers, no matter how strong their individual contributions. No higher-level program assessment, no matter how carefully structured or replete with data, can improve without the input of classroom teachers, including a coordinated, self-conscious, and collaborative implementation of pedagogical strategies.

Two fictitious cases—one involving a technical communication program and one involving a department of chemical engineering—will serve to illustrate this principle. In each case, program leaders aspire to help students achieve outcomes they hold to be important within the scope of their curriculum. But just as they are struggling to define and oversee an assessment process from the outside in, teachers within each curriculum are also providing valuable insights—from course- and classroom-based perspectives—that in turn reinform the assessment process from the inside out.

FROM THE OUTSIDE IN

The formulation of learning outcomes for programs of study offers us a way, as Hundleby and Allen (this volume) put it, to *begin*. Articulating outcomes requires us to reflect on what we want students to know, acquire, or be able to do as a result of what we provide. In so doing, we ask both philosophical and practical questions about our disciplines, their goals, and their roles in a complex intellectual culture. We are compelled to ask whether the expectations we draft are reasonable, realizable, measurable. Can we see them? What do we really mean by them? These discussions often energize groups of instructors and program administrators because they tap into their tacit experience, sometimes years of it, drawing out assumptions about what students are or are not capable of doing and what we aspire to help them learn. At their deepest and most productive, they can also help us stand back from our disciplines and see them as constantly evolving domains of intellect and activity, intertwined with other areas of knowledge and in need of regular curricular revision. As our two assessment cases illustrate, the process often begins with tacitly held beliefs about students' work. The resulting outcomes may represent a halting and even less-than-fully theorized beginning; but it is crucial that the outcomes emerge from the discussions and negotiations of the teachers and administrators within the program.

Case 1: *Administrators of a Technical Communication program housed in an English department and providing courses in the service of many scientific and professional majors across campus are working on their curriculum. Knowing the importance of understanding and distilling information from scientific journals, they have decided to include the outcome that students will "show that they can effectively summarize information from professional articles." The importance of this outcome is evident in much of the technical writing literature, in their experience working with instructors in a range of technical disciplines, and in their sense of broader expectations for student writing in academia. They believe that this ability can be described and demonstrated, and that it varies mainly in matters of personae (whether the writer is "in" the summary) and appeals to audience (how much "translating" needs to be done relative to the reader's level of knowledge and expertise). As they talk, they are exploring questions at the heart of assessment: what is it? How do we describe it? How does it vary? How important is it?*

Case 2: *Several leaders of a chemical engineering department have gathered to create outcomes for student learning with a strong focus on communication abilities. As is typical, their discussion moves between the emerging draft of their outcomes statement and that "felt sense" of what they think students ought to be leaving with as majors. When the matter comes up, they agree that chemical engineering majors often make errors in writing, even as seniors: lapses in* *punctuation, slips of grammar, awkward constructions, which betray problems with control at the surface of their prose. As their discussion weaves through anecdote and evidence, head-shaking generalizations and memories of specific students or texts, they too are asking, what is it? How do we describe it? How does it vary? How important is it? On matters of linguistic felicity, such questions are not particularly challenging. Beyond simple negotiations of "what it is," no one in this group disagrees that the "it" is important. It varies by student, with the nonnative speakers of English representing a special challenge. But it strikes them as a reasonable outcome: chemical engineering majors will demonstrate that they can write academic papers and reports with only occasional errors of grammar, punctuation, and other surface-level matters.*

As powerful as the process of articulating outcomes may be in helping a program to reach consensus about what matters in student learning, the outcomes remain inert—and unchecked—until they can be assessed and reconsidered in light of that assessment. Collecting data on outcomes makes us ask how and where they are manifested, yielding highly productive discussions about teaching and learning. Where can information on particular outcomes be obtained? At what point should that information be collected? How should it be assessed? Should information be sampled from a range of courses across the curriculum, or should the context be more targeted? What sorts of products will best show whether the outcomes are being achieved? Should assessment be direct—for example, from written products—or should other, indirect measures be used, such as surveys of students, faculty, alumni, or employers? This process of choosing the contexts, timing, and tools of assessment moves program administrators from the wishful thinking that outcomes represent to concerted efforts to gauge their manifestation (or lack thereof) in the abilities of students at some point in time. It represents a challenging but indispensable part of the assessment process, and its results can provide focus for course improvement, give instructors and administrators direction for their work, and ultimately yield positive change in the curriculum. As our two cases unfold, we also see that formulating an assessment plan does not always lead to a reconsideration of problematic outcomes.

Case 1 (continued): *In order to assess students' abilities to summarize professional articles, leaders of the technical communication program decide to sample identical summary assignments from two sections of the same Technical Communication course required of majors in a number of disciplines. With the cooperation of the two instructors, they collect copies of summaries from early and late in the course, in a kind of pre-/post design. Using a primary-trait rubric of desired features, they score the sample summaries, reaching a high degree of interrater reliability. Using these methods, they hope to assess an important outcome of their curriculum in a systematic and representative way.*

Case 2 (continued): *As they consider how best to assess the outcome focusing on surface correctness, the members of the chemical engineering department decide to collect a random sample of papers from majors in four required courses. In cooperation with the registrar's office, department leaders randomly select the names of a subset of students in each of the four courses. Instructors are asked to photocopy a paper—to be written in response to formal assignments of at least 3–4 pages—from each of the designated students. The submitted papers are then subjected to an error count, based on a set of predetermined errors deemed both important and representative of students' control of surface features in their prose. Each paper's errors are tabulated by two trained reviewers; significant disagreements between the two are resolved by a third reader. Using these methods, the department hopes to assess one of its outcomes in a systematic and representative way.*

In any assessment program, the most challenging but important question arises when the data have come in and been analyzed—that moment when one or more outcomes stand the test of assessment reality: *now what?* This question is about the consequences of the assessment, the implications of a plan of action, of change, of getting closer to what's desired in student achievement. Since it's rare for outcomes to be initially measured at a completely satisfying rate, there is almost always some work to do. What do we do to improve? Where do we go from here? In grappling with decisions of implementation and curricular change, the conversation turns from "What do we expect or hope for?" and "How do we find out how we're doing?" to "Now that we know where we are, what do we do?" Because that conversation usually involves changes in curriculum, changes in teachers' methodologies (or entire ideologies of teaching and learning), and changes in the status quo, it is often among the most politically and interpersonally charged of all the processes in the assessment of learning. As our cases continue, we also see the ways in which values associated with writing initially influence considerations of action.

Case 1 (continued): *After carefully assessing sample summaries from two sections of their popular Technical Communication course, program leaders discover, to their chagrin, that the scores from both sections of the course call for significant attention to this genre of writing in students' experience. In addition, the scores from one section show almost no improvement between the early and late samples, while those from the other section show significant improvement, even though they do not rise to an acceptable level. This suggests to the program leaders a possible relationship between instruction and performance. As they reflect on these results, the program leaders come face-to-face with the "now what?" question. Three possibilities come to mind early on: First, it is tempting for them to think about working with the teacher whose students yielded the lower* aggregate scores on improvement in summary writing. However, they reject this course of action immediately, knowing that it risks creating the perception that program assessment is really about individual teacher assessment (which it's not), and in turn this could create paranoia or subvert further efforts to collect useful information and gain the cooperation of instructors. Second, they could begin an initiative in which instruction in summarizing is integrated into all courses in the Technical Communication curriculum. This could create shared knowledge and strategies across instructors and sections, and lead to more possibilities for further assessment, including looking at summary ability more developmentally. Third, they could extend and refine the assessment itself, gathering more information about the genres of summary, the specific deficiencies students experience in meeting the program's outcome, and the instructional conditions under which students write and learn about summarizing. This information, in turn, could lead to the modification of this and other outcomes.

Case 2 (continued): *The assessment of chemical engineering majors' writing has yielded a rather high percentage of surface errors as a percentage of total text. ChemE majors, it turns out, are just not good at controlling their syntax and grammar, punctuation, referencing, even spelling. As the department leaders muse over the disappointing results, they must now turn to action.* Now what? What should be done in the curriculum, or beyond, to rectify this empirically demonstrated lapse in the achievement of an important writing outcome?

As they consider their options, six solutions emerge. First, the department could ask that all instructors require any students deemed deficient in this outcome to go to the Writing Center and get tutorial help, which is documented and communicated to the referring teachers and subsequently to the department. As a second alternative, the department could communicate its findings to the first-year composition program and ask them why Chemical Engineering majors, most of whom have taken the required freshman sequence, are still deficient as juniors and seniors in their control of surface features, and perhaps even demand improvements in the program. Third, a new departmental policy could be established that would "send a strong message" to students deficient in these skills by failing them for more than a set number of errors of various kinds in their formal writing. Fourth, a special grammar course could be created in the Chemical Engineering department and either required of all majors or required of those majors who are deemed to be deficient. Fifth, all instructors in the department could be required to brush up on basic grammatical knowledge and spend time in their own classes teaching some of the rules to their students. Sixth, barrier tests could be set up for the end of the junior year to assess students' control of surface correctness, and those not passing the test could be required to take an additional course offered by the English department or otherwise demonstrate their improved abilities through subsequent tests.

As these two cases of the assessment cycle illustrate, considering programmatic solutions to problems identified through systematic assessment reminds us that assessment is *almost always ideological*. That is, the entire process, from articulating outcomes to going about collecting data on them to asking "*now what?*" when the results come in, depends on collective agreements about what matters and what to do about it. At base, assessment is, as Norbert Elliott points out (this volume), ontologically and socially constructed.

In the case of the chemical engineering department, the familiar ideology that links "good writing" with correctness of grammar and other surface features of text initially influences the department's construction of outcomes and assessment plans. The decision to focus general learning outcomes for Chemical Engineering majors so strongly on the demonstration of correctness in writing seems, from the perspective of many administrators and instructors in both writing programs and engineering departments, misguided. To divorce this outcome from other centrally important considerations of writing in engineering, such as the representation of phases in project work or the ability to translate sophisticated scientific prose into persuasive, understandable material for marketing teams or management, seems wrong-headed. To spend significant assessment resources, and to send a message *through* the assessment process, that hooks and dots, spelling, and the location of prepositions are among the most important abilities Chemical Engineering majors can aspire to, seems, well, just silly.

However, in the absence of guidance from other experts—guidance of the kind described by Michael Carter in his collaboration with an engineering program (this volume; see also Anson, Carter, Dannels, & Rust, 2003)—our emblematic Chemical Engineering department has all the right in the world to shape its assessment process around what may appear to be somewhat unprincipled or untheorized approaches and focuses. Often this is what happens in the *beginning*. But if assessment becomes fully and thoughtfully recursive, it is self-rectifying, leading to reconsideration not only of the curriculum or the methods used to get information about it but the very outcomes that are the progenitors of the assessment program. Consider, for example, what happens when the Chemical Engineering department's leaders seriously reflect on each of the six solutions they have cooked up to improve demonstration of the correctness outcome. Blaming the composition program does nothing, though meeting with its leaders is a good start to learn what might be expected from the required sequence and to hear from experts about the problem of the transfer of ability across diverse contexts, genres, and audiences. Creating a grammar course in chemical engineering makes little sense, even if there are resources to do so—and who will teach it? Sending so many students to the Writing Center, as if its role is always corrective and remedial, will meet with resistance. Failing students for small infelicities is irresponsible. Creating barrier tests for a microportion of all that should define a well-prepared Chemical Engineering major skews priorities. The myriad problems raised by these proposed courses of action do not point to a lack of adequate resources to address a perceived need; they point *backwards*, or perhaps *forwards*, through the loop, to the serious flaw of that perceived need in the department's ideology of student learning and achievement. This is not to suggest that correctness and the adherence to the conventions of written discourse should be ignored; it's to imagine that the Chemical Engineering department can weigh such an outcome more thoughtfully relative to the many other intellectual and disciplinary expectations they have of their curriculum and their majors, including those influenced by developing technologies and literacies.

The first case presents a similar set of problems, although the outcome under consideration suggests somewhat more careful alignment with the goals and nature of the Technical Communication curriculum and its role in general education—perhaps also a beginning, but a beginning that is farther along. Here, though, we see greater sensitivity not only to possible implementation strategies, with a concern for the integrity of the process and an interest in keeping the involvement of instructors, but to the entire assessment cycle itself. The program leaders are already poised to reconsider both the way they are looking at summary data and, if necessary, the outcome they created.

As described, the first case gives some confidence that program leaders will continue to modify and adjust their outcomes cycle just as they engage in the process. In the case of the Chemical Engineering department, we can only hope that initially groping their way through this difficult process might lead department leaders to similar kinds of changes. But how can we be sure?

When the process of assessment is orchestrated entirely from the outside in, by administrators and program leaders; when its outcomes and data-gathering decisions are formulated from without; and when the implications of assessment are imposed on curricula and teachers without their input, there is a much greater chance for the failure of the assessment program to create positive change manifested in improved student learning. Assessment also needs to work from the inside out.

FROM THE INSIDE OUT

Program assessment is often kept at arm's length from course assessment, which focuses on the improvement of a curricular offering or a specific section of it, or the kinds of assessment instructors are encouraged to engage in under the provisions of "reflective practice" (Schön, 1983) or, more generally, the scholarship of teaching and learning (Hutchings & Shulman, 1999). Program assessment is "the systematic and ongoing method of gathering, analyzing and using information from various sources about a *program* and measuring *program outcomes* in order to improve student learning" (Selim & Pet-Armacost, 2004, p. 10; emphasis added), whereas course assessment refers to the collection of information that can help improve the structure, design, teaching, and evaluation of a specific *course*. The closer one moves in on a specific course, the stranger it becomes to assess it from the outside, without the involvement of those who actually teach it. Ideally, course-level assessment can provide valuable information for broader program-related goals, as illustrated in Doreen Starke-Meyerring and Deborah Andrews' assessment of their intercultural virtually linked courses and Deborah Bosley's similar intercultural collaboration around one assignment in a technical writing course (this volume).

Working from the inside out means helping instructors to engage in the kinds of assessment practices that are crucial to the ongoing improvement of teaching and the cumulative development of teaching expertise. Although the assessment cycle is quite similar, individual teachers have more opportunities to engage in situated, authentic assessment, because they can collect and analyze information and use it to make decisions about their teaching in an unobtrusive way, completely grounded in the daily work of their instruction. Furthermore, their options for assessment can range from informal reflective practice—making observations and hypotheses in the classroom, reflecting on them, actively experimenting with new methods and strategies, and reflecting anew—to "action research" (Kemmis & McTaggart, 1988; Sagor, 1993) and more formal investigations advocated in the literature on the scholarship of teaching and learning (Cross & Steadman, 1996).

Case 1 (continued): *In our representative technical communication program, Lotte Hagstrom has been teaching an upper-level course in technical writing populated by majors in various professional and technical disciplines. Early in the course, an important assignment on which larger papers are later scaffolded is a summary of a reading from the professional literature in the student's chosen field. This assignment also supports one of the program's stated outcomes. In working through this assignment, Lotte has noticed that some students tend to summarize at inappropriate levels of detail, unable to synthesize the information from their articles into a brief but coherent representation of the whole. As she moves around the small discussion groups she has set up for students to share the information in their readings prior to writing their summaries, she notices that some students lack confidence talking about the material they have read. Something in their reading process seems to be missing. Reflecting on these lapses, she decides to experiment with some new classroom strategies to help students to critically analyze their chosen texts.*

Case 2 (continued): *Rich Denizo routinely teaches CHE 306, "Chemical Reactor Design," in our sample Chemical Engineering department. This required course involves 20 pages of formal writing, which Rich usually divides between three open-ended problem analyses and a longer paper involving the creation and description of models. Over the past few years, he has noticed that* *his students continue to have problems controlling grammar and surface mechanics in their formal writing. Admonishing the students—in class and in the syllabus—hadn't made much difference. When he doubled the points deducted for such errors, he became embroiled in indecisions and regrets in his grading, feeling unfairly lax in some cases and overly vindictive in others. He soon reverted to his previous grading scheme but remained vexed by the problem, increasingly so since department administrators had been focusing sharply on the problem of surface errors and toying with outcomes assessment of the same.*

In studying students' papers more carefully, he now notices that what he usually lumped together in the category of "error" may be more complicated than he'd thought. Some errors appear to be simple slips and oversights, showing up and disappearing in the same paper. The students must know the correct forms, because not every case is erroneous. In fact, almost half the errors are those the student don't make every time. Reflecting on this observation, Rich decides to experiment with a peer strategy he learned about in a faculty-development workshop a couple of years ago.

Unlike program-level assessment, which often looks to broad curricular changes when its outcomes are not fully realized, course- and classroom-level assessment raises questions not just about requirements and grading criteria but about *support for student learning*. When students fail to learn something fully or can't demonstrate that they have acquired some skill or knowledge, instructors must finally ask what they are (or are not) doing that relates causally to their observation. Support lies at the vector where assessment flips from outside to inside, inside to outside. Without it, students won't learn, and no assessment of that learning, no matter how robust or meticulous, will make a bit of difference for all the time and energy expended. Yet as Jablonski and Nagelhout demonstrate in their contribution to this volume, support for learning requires support for teaching, in the form of faculty-development programs, electronic technologies, collaboration among instructors and administrators, and a strong culture of teaching and learning that provides incentives and rewards for engaging in the ongoing challenges of working toward excellence.

Case 1 (continued): *Lotte Hagstrom's classroom-based experiments have led her to develop instructional materials designed to help students read more carefully and critically prior to writing their summaries. These materials include Web-based strategies for prereading and postreading; in-class small and large group work focusing on synthesizing main ideas from readings at appropriate levels; and a new mid-draft intervention based on a set of generic reading questions. After implementing these strategies, Lotte notices a sharp improvement in the quality of students' summaries, and especially in their appropriate level of detail.*

In conveying the results of this work to the program administrators, Lotte also affects the nature of the assessment program. *Clearly, simply requiring more summarizing across the technical communication curriculum may not be as effective as introducing instructional methods that focus not so much on students' writing but on their reading. As these new orientations and strategies are incorporated into the curriculum, program leaders can continue to sample student work and gauge the distance between performance and their important outcome involving the synthesis of professional literature.*

Case 2 (continued): *Although Rich Denizo has always been somewhat skeptical about the process of peer review, he decides to experiment with it during what he calls the "final editorial clean-up" of his students' formal writing. Students share their near-final drafts electronically in groups of three between classes. For homework, they must identify as many surface problems as possible in their peers' drafts and also in their own, then bring printouts of the drafts to class with the identified errors. In 25 minutes during class, they share the problems they have identified, comparing peers' responses to their own editorial corrections. Any ambiguous or unresolved problems need to be researched using a handbook or other appropriate resource. After incorporating this strategy into his CHE 306 class, Rich immediately notices that the proportion of surface errors in students' papers decreases significantly. Those that remain appear to be much more conceptual—part of each student's particular grammar—and not just simple slips. But as he considers how many errors the authors themselves were able to identify in their own papers, it occurs to him that the solution isn't so much in an editorial review session but in preventing students from writing papers at the last minute, when sowing the seeds of writing at such a panicked and frenzied pace yields a bumper crop of weeds amidst the rows of ideas. A full peer-review process, he reasons, builds in a prescribed pace, with opportunities for revision, editing, and proofreading along the way. He is newly inspired to rethink his objections to multiple drafts and review sessions.*

In conveying the results of these informal investigations to his department leaders, Rich influences the entire assessment process, *casting a portion of the perceived error problem in a new light. Chemical Engineering majors are not devoid of knowledge at the surface of their prose; they are just writing under the wrong conditions. Classroom strategies like peer review will help, supplemented by a strong dose of instruction in how to avoid procrastination, plan writing in stages, and take time to edit for surface problems. Furthermore, in this light, the avoidance of error no longer seems like an A-level outcome, perhaps better subordinated under a new, general outcome concerning students' overall abilities to plan, draft, revise, and edit their papers effectively.*

TOWARD CONTINUOUS ASSESSMENT

Assessment runs in a seamless, undivided but unique loop involving those working at different levels of a program. The contributions to this collection, with their focus on a range of assessment programs, practices, and philosophies in the area of technical communication, demonstrate the need for multilevel articulation and collaboration in the effort to improve educational practice. The challenges to such multiple articulations and collaboration are many, as Cook and Zachry point out in "Politics, Programmatic Self-Assessment, and the Challenge of Cultural Change" (this volume), but they are not insurmountable.

Although the structure of the collection suggests a kind of linearity of design—primary articles followed by comment and response pieces—the contributions also speak to each other across their own boundaries, covering a range of topics and focuses at both the programmatic and course levels and through different genres and techniques. Questions raised in one piece echo back in others, but throughout there is a sense of continuous exploration, and an accepting attitude that, once we begin the process of program assessment, we step into a loop that has no definable end.

The dialogue set up among the contributions also suggests that we need to think more fully about *interprogram* assessment, about connecting the assessment of different programs to each other. If program assessments are divided from each other, they lose their potential for mutual support and clarity of purpose. Our fictitious cases, for example, ultimately demonstrate useful assessment activity within their own curricular domains. But consider what happens when they decide to have a new conversation about how their two programs relate to each other in the service of student development.

Cases 1 and 2 (continued): *Leaders of the Technical Communication program have been slowly but systematically meeting with administrators in some of the departments whose majors are most populous in the Tech Comm. courses they offer. By sharing each other's outcomes and assessment data, they begin to see patterns and relationships, redundancies and gaps, in their preparation. What is going on in the Tech Comm. course that the Chemical Engineering courses can assume? What is going on in the Chemical Engineering courses that needs special preparation or focus in the Tech Comm. curriculum? How can two different and haphazardly-related sets of outcomes be revised to reflect a better understanding of student development in communication and a clearer scaffolding of course experiences?*

We must also remember that outcomes assessment represents only a portion of all the aspirations of teachers and students, not all of which are measurable and some of which are not, at least directly, even teachable. Outcomes assessment is a *beginning*, based on the principles of clarity, articulation, measurement, and systematic implementation. It runs the risk of perpetuating a paradigm of higher education inherited from industrial models of efficiency and mass production—a kind of assembly line with a lot of quality control. The full input of teachers helps to guard against a pedagogy driven, not guided, by an interest in discrete sets of skills or abilities, or by measures of accountability instead of a passion for continuous improvement based on formative rather than summative feedback.

Full articulation of all participants, particularly students, also allows for assessment not to push into the background anything that is not directly measurable. At an accredited Native American university I once visited, the cultural values that drive academics there include a strong focus on teaching to the "heart" as well as the intellect. Although impossible to turn into a "teachable, measurable" outcome, this and other cultural and educational values are present in much of what goes on at the institution—in everything from its mission statement to its course instruction to its sports programs and student activities. Similarly, aspiring to help students develop a "lifelong love" of a subject, or to turn them into "ethical leaders," or to give them "intellectual curiosity" are goals that should not fade away in the effort to study, track, map, gauge, or assess those elements of learning that *are* measurable. Sometimes institutional values can be incorporated into program objectives, "increasing the intellectual strength of the curricula and extracurricula we offer students," as Paul Anderson and Jo Allen show in their conversation in this volume. But not everything is, or should be, about assessment.

REFERENCES

Anson, C., Carter, M., Dannels, D., & Rust, J. (2003). Mutual support: CAC programs and institutional improvement in undergraduate education. *Journal of Language and Learning Across the Disciplines. Special Issue, 6*(3), 26–38. Retrieved on May 5, 2009 from http://wac.colostate.edu/llad/v6n3/anson.pdf

Cross, K., & Steadman, M. (1996). *Classroom research: Implementing the scholarship of teaching.* San Francisco: Jossey-Bass.

Hutchings, P., & Shulman, L. (1999). The scholarship of teaching: New elaborations, new developments. *Change, 31*(5), 10-15.

Kemmis, S., & McTaggart, E. (Eds.). (1988). *The action research planner.* Melbourne: Deakin University.

Nichols, J. O. (1995). *Assessment case studies: Common issues in implementation with various campus approaches to resolution.* New York: Agathon.

Sagor, R. (1993). *How to conduct collaborative action research.* Alexandria, VA: Association for Supervision and Curriculum Development.

Schön, D. (1983). *The reflective practitioner: How professionals think in action.* London: Temple Smith.

Selim, B. R., & Pet-Armacost, J. (2004). *Program assessment handbook: Guidelines for planning and implementing quality enhancement efforts of program and student learning outcomes.* Tampa, FL: University of Central Florida.

CHAPTER 2

Assessing Technical Communication: A Conceptual History

Norbert Elliot
New Jersey Institute of Technology

As children of Hegel, we follow the dialectic metaphor of progress wherever it may lead. Interplay produces innovation. We thrive in a present that is inherently superior to the past, anticipate a future immanently predictable because it, too, will behave in accordance with the dialectic. The teleological fix is in. As children of Derrida, conversely, we follow the temporal metaphor of contingency wherever it may lead. Contextuality produces criticality. We thrive in a present that is inherently disconnected, expect a future immanently articulated because it, too, will open itself in accordance with the contingency. The teleological fix is in. Whether the dropper of bread crumbs is *Geistes* or *aporia*, we find ourselves walking down a comprehensible road. If we tend toward the buttoned-down (*"Whatever is, is right"*), we seek solace in structure; if we tend toward the open-necked (*"Still in the published city but not yet overtaken by a new form of despair, I ask the diagram: is it the foretaste of pain it might easily be?"*), we seek solace in contexts.

The interplay of modernism and postmodernism engages and informs us as we undertake the assessment of technical communication. Establishing viewpoint provides insight into both the social circumstance of an idea and the way that idea reflects the social circumstance within which it is situated. Examining heuristic value and cultural embeddedness allows us to understand that paradigms shuffle. In this chapter—a call, ultimately, to embrace informed aims of strategic assessment—I begin with an examination of the Defense Activity for Non-Traditional Education Support (DANTES) Subject Standardized Test of Technical Writing as an icon of the modernist project. I will then turn to the deeply contextualized work of Lee Odell and Dixie Goswami, an implicit rejection of the modernist project, in their study of writing in a nonacademic setting. The pressure of institutional and communal forces is enormous where accountability is concerned, and it is time to understand these forces, identify our traditions, and make up our minds about the directions we need to take in the assessment of technical communication. But to appreciate this contemporary part of the story, we must first turn to an earlier episode.

MODERNISM: STRUCTURALIST TESTING

The contributions to American material life made after the Civil War reveal a singular truth: engineering manufactured the mesh holding together the nation's daily life. When T. A. Rickard wrote in his third edition of *Technical Writing* that "the purpose of writing is to convey ideas, and that ideas can not be conveyed successfully in defiance of technique," few in 1931 were in a position to believe otherwise (p. 4). For better or worse, the artifacts of technique were everywhere. The metaphor of a system—classificatory, measurable, accountable—was uniformly adopted in America, as David Harvey has shown, yielding a coherent schema of reproduction (1989, p. 122). To assist in training engineers to communicate, an essential part of such systematization, teachers of technical writing came to hold a recognized place in the academy before World War II (Connors, 1982). As Jeff Todd has observed, "Through technical writing, engineers were able to stabilize management's control over labor, as well as the subjection of management—and themselves—to the system" (2003, p. 76). By the time Dwight David Eisenhower became the nation's 34th president, courses in technical communication were increasingly becoming part of the college curriculum.

In similar fashion, social measurement had begun to take on a solid presence in everyday life. Revising the tests of Alfred Binet, Stanford University's Lewis M. Terman developed the Intelligence Quotient. As president of the American Psychological Association in 1917, Robert M. Yerkes (1921) lobbied the surgeon general of the Army to utilize intelligence testing in order to evaluate the mental abilities of soldiers. With tests developed for recruits who were either literate (Army Alpha) or illiterate (Army Beta), Yerkes and his 16 psychologists had tested 1,555,256 soldiers when the examinations ended shortly after Armistice Day. Among those psychologists was Carl Campbell Brigham. As a member of the faculty at Princeton University after World War I, Brigham used the data gathered in the Army testing to publish *A Study of American Intelligence* (1923). There, the young assistant professor of psychology identified statistically significant differences in the Army tests among immigrant groups and, thus, reified the past restrictive immigrant policies of 1917 and licensed the future of American eugenics.

Social measurement, with its basis of integrated concepts of intelligence and literacy, had become, along with engineering, a significant force in everyday life. When Brigham turned his attention to the assessment of academic ability, his new Scholastic Aptitude Test became an efficient way to judge individual potential—a scientific answer to Harvard University President's Charles W. Eliot's belief, voiced a generation before, that admissions tests prevent "a waste of instruction upon incompetent persons" (1869, p. lxiii). By the time Eisenhower authorized the interstate highway system, the College Entrance Examination Board—the sponsor of Brigham's test—had spun off its research division and fostered the creation of the Educational Testing Service (ETS). "For education as a whole," ETS President Henry Chauncey wrote in his 1957–1958 *Annual Report*, "a new era seems to be clearly at hand, in view of the tremendous increase of public awareness and appreciation of the fundamental importance of education to the welfare of each individual, the progress of our rapidly changing society, and the security of the nation" (p. 17). As that organization became firmly entrenched within the nation's educational system, modernism's coherent schema of reproduction was established, a key factor in both the pursuit and evaluation of success.

The definition of technical writing that would emerge during this period is congruent with the faith in immanent predictability that marks the era. As Donald H. Cunningham recalled his experiences from the mid-to-late 1960s through the 1970s, "The field lacked established forums for sharing of information, and people sought help in an ad hoc way" (2004, p. 122). Definitions were, at first, hard to find in a classroom environment in which instruction was, Cunningham notes, carried out by semi-interested participants who used literary examples such as Thoreau's observations about black and red ants as exemplary readings. Certainty was found in that which was familiar to instructors trained to believe in the universal lessons of literary analysis gained by that most scientific of critical techniques, the new criticism (Eagleton, 1996, pp. 37–46). The textbooks of the day were Gordon H. Mills and John A. Walter's *Technical Writing* (1954) and *Technical and Professional Writing: A Practical Anthology* (1963), edited by Herman A. Estrin. The emerging definitions of technical writing were based on format, style, and content. As Walter found in classifying technical documents that he had reviewed, each text had precise rhetorical modes aimed at a specific audience, a unique vocabulary, and a demonstrable technical content (1977). John S. Harris (1978) declared similarly that technical writing was to be defined as "the rhetoric of the scientific method," firmly enrolling the field into the modernist project (p. 12). As Harvey (1989) describes the broad cultural process,

> The idea was to use the accumulation of knowledge generated by many individuals working freely and creatively for the pursuit of human eman-cipation and the enrichment of daily life. The scientific domination of nature promised freedom from scarcity, want, and the arbitrariness of natural calamity. The development of rational forms of social organization and rational modes of thought promised liberation from the irrationalities of

myth, religion, superstition, release from the arbitrary use of power as well as from the dark side of our own human natures. Only through such a project could the universal, eternal, and the immutable qualities of all humanity be revealed (p. 12).

When the Association of Teachers of Technical Writing (ATTW) was founded by Cunningham and his colleagues in 1973, its aim was to reinforce the project at hand, a modernist conception of science. Technical writing was effective to the degree that it was transparent: written language existed on the page solely to capture the logic of an established empirical process. To make it new, as Ezra Pound had said in 1934, was, in the case of the project of technical writing, to serve scientific innovation. It is no surprise, then, that what held true in defining the construct of technical writing held true for its assessment.

Emblematic of a modernist view of evaluation was the 1974 creation of the Defense Activity for Non-Traditional Education Support (DANTES) Subject Standardized Tests. Each of the DANTES Subject Standardized Tests (DSSTs in the areas of applied technology, business, foreign language, humanities, mathematics, social science, and physical science) was to be understood as a subject-matter achievement test "patterned according to the typical offerings of academic and vocational institutions" (American Council on Education [ACE], 1987, p. 61). Each test carried American Council on Education credit recommendations; therefore, a candidate who succeeded on the test could apply for college or university credit because the ACE—founded in 1918 and serving as a unifying forum and, in this case, credentialing agent for postsecondary education—had accepted the tests as valid. In 1982 the contract to develop the tests was awarded to the ETS. The tests were administered to those in the armed forces at a rate of 15,000 to 20,000 examinations per year from 1982 to 1986. Concurrently, the program was made available to civilians in 1983. By 1985—the launch date of the test in Technical Writing (Form SE-820, situated within the Applied Technology test area alongside the test for Basic Automotive Service)—approximately 500 civilians took DSSTs (ETS, 1985). The teleological fix was in. The system, able to achieve massive growth, was in place. The vehicle for that system, the test itself, was designed to facilitate such growth.

During the spring of that year, 45 colleges and universities participated in establishing national norms for the DANTES test in Technical Writing. The statistical data that would summarize the performance of the new test as taken by 687 students who were completing, or had recently completed, an elementary university or college course in technical writing. The content of the Technical Writing test was based on six topics commonly taught in the classroom, and a percentage of the test was allocated to each topic: theory and practice of technical writing (4%), modes of technical writing (27%), technical documents (21%), parts of formal documents (7%), organizing technical information (13%), and technical editing (28%). Sample questions, suggesting a remorseless certainty regarding genre, are given below from the ETS Fact Sheet/Study Guide (1985, p. 3):

4. The main difference between proposals and many other technical documents is that proposals are
   (A) long and formal
   (B) written by committee
   (C) overly persuasive
   (D) presented orally as well as in writing

5. Laboratory reports customarily contain all of the following sections EXCEPT
   (A) Materials and Methods
   (B) Results
   (C) Discussion
   (D) Recommendations

There were also written passages to be organized for coherence, such as a passage of disjointed sentences about rabbits building up fat in the walls of their arteries and a series of multiple-choice questions asking which sentence in the passage should appear last.

Additionally, there were sentences offered for revision (e.g., the sentence, "After heating for twelve minutes, the sample was dried" and a series of multiple-choice questions, each rephrasing that dangling modifier). The Technical Writing test also required that each student write a short essay on one of a series of provided technical topics, but the essay—its open response answers a known impediment to efficiency—would not be scored. Because all DSSTs had the same cut score, and since all the tests were nationally normed in the classroom, colleges could reason that a passing score on the test would indicate that the candidate had obtained an acceptable level of subject knowledge, immutably defined as the equivalent of the grade of C, earned by a portion of the 687 technical writing students in the norming sample. In the 1985 ETS Fact Sheet/Study Guide containing the content and sample-question information, Reference Publications were also provided: *Technical Writing: Principles and Forms* (Andrews & Blickle, 1978), *Handbook of Technical Writing* (Brusaw, Alred, & Oliu, 1976), *Reporting Technical Information* (Houp & Pearsall, 1968), *Technical Writing* (Lannon, 1982), *Technical Writing* (Mills & Walter, 1954), *Scientific and Technical Writing* (Sandman, Klompus, & Yarrison, 1985), and *Basic Technical Writing* (Wesiman, 1985). No dates were given for any of the publications in the Fact Sheet/Study Guide, their absence suggesting that the test developers possessed a firmly embedded definition of universal technical writing qualities that were immutable. This enduring sense of the construct validity of technical writing, identified clearly and employed judiciously, could then be efficiently operationalized through limited response (multiple-choice) questions.

By 1989 the DANTES program had seven categories of Subject Standardized Tests, as well as a category of Applied Technology in which the new Technical Writing test appeared, along with a test of Principles of Refrigeration Technology. The ETS 1991 Fact Sheet/Study Guide revealed no changes in the test content, item format, or reference publications, although students could indicate, based on college requirements, whether or not they wished to write the nonscored essay. In 1993 the Technical Writing test was normed a second time (Schwager, Gall, & Thomas, 1995). The new norm process employed a national sample of 1,257 students enrolled in technical writing courses offered in 68 public and private universities, 4-year colleges, and junior colleges. The cut score remained the scaled score corresponding to the mean score of the students who received a technical writing course grade of C. The established cut score, yet another feature of modernist structuralism, thus remorselessly welded the validity of the test to the validity of the course. Under a 4-member test development committee (Warren Buitendorp of Montgomery College [MD], Naomi Given of Middlesex County College [NJ], Ralph J. Patrick of Cheyney University of Pennsylvania, and Rita Reaves of East Carolina University), the content categories had efficiently decreased to four: theory and practice of technical writing (10%); purpose, content, and organizational patterns of common types of technical documents (31%); elements of various technical reports (31%); and technical editing (28%). In the 1994 Fact Sheet/Study Guide publicizing the newly normed test, the item types remained the same, and the short essay remained elective and nonscored. The Reference Publications retained Brusaw, Alred, and Oliu's *Handbook of Technical Writing* (4th ed.), Lannon's *Technical Writing* (5th ed.), and Mills and Walter's *Technical Writing* (5th ed.). Edition dates were now provided. Added was a list of technical writing journals: *Journal of Technical Writing and Communication*, *Technical Communication*, and *The Technical Writing Teacher*. A Fact Sheet/Study Guide published in 1998—as well as one published in 2002 (by the Chauncey Group, then the test owner)—revealed no change in the test content, sample questions, reference publications, or recommended journals. In 2004 Thomson Prometric acquired the Chauncey Group International, and in 2007 Thomson Prometric was acquired by ETS—the original sponsor of the test. Regardless of owner, little has changed in 22 years. *The Technical Writing Teacher* is still recommended to help candidates prepare for the test, even though that journal had become *Technical Communication Quarterly* 17 years earlier.

As it presently exists, the DSST in Technical Writing remains a modernist metaphor. Manifest is the definition of technical writing as Harris had defined it in 1978: technical writing remains the rhetoric of the scientific method. The Thomson Prometric Fact Sheet tell us so: knowledge about facts and terms, understanding of concepts and principles, and the ability to apply knowledge to specific problems and situations all endure, captured efficiently by multiple-choice questions (2007, p. 2). In that there may remain some sort of untidy complexity in play in the construct of technical writing, there is that optional essay "portion," but candidates are reminded that "Thomson Prometric will not score the essay" (2007, p. 1). The inefficiency in scoring remains, passed on to colleges and universities. Assumed to be a transparent medium, the Technical Writing test manifests a rational existence steadfastly devoted to capturing and reproducing the logic of science. The construct validity of the test, informed by elemental definitions, can—at its most mature—achieve only the slightest change, because the construct must reinforce the requirements of an immutable scientific enterprise. Attention to the theory and practice of technical writing rises from 4% in 1985 to 10% in 1994, and attention to the aims of technical writing—purpose and content—displaces attention to modes. Yet that is all. All fact sheets and study guides remain nearly identical in validity orientation. All are mirrors that reflect a universe steadfastly understood to be objective.

Logically, the test format itself reflects a construct that always has the potential to supply needed content validity, because there exists an objective series of rules for writers to follow in the service of science. Since it is unnecessary, the open-response format of a writing sample is an unwelcome externality, its scoring a burden passed on to others. The test development process, in the service of both the immutable nature of science and transparent technical writing enterprise, goes on, indifferent, seeking only to update itself, on occasion, with a new national norm. The only responsibility is to reflect the engineered society within which the test is situated.

POSTMODERNISM:
CONTEXTUAL ASSESSMENT

As Ihab Hassan long ago pointed out, modernism and postmodernism are not separated by an Iron Curtain or a Chinese Wall. History is a palimpsest, he recalls, and the idea of a historical period belongs more to historians than to history (p. 121). Taking up postmodernism as an alternative perspective allows us, as Hassan proposes, to embrace both continuity and discontinuity as we replace modernist concepts (closure, purpose, design, distance, centering, selection, *Grande Historie*, metaphysics, determinacy, transcendence) with their opposites (openness, play, chance, participation, dispersal, combination, *petit historie*, irony, indeterminacy, immanence) (1985, pp. 123–124). Recognition of the need for the profession of technical writing to invent opportunities for cultural power and authority, to cross boundaries, and to assert the value of citizens who are, ultimately, influenced by the presence of technology in their lives (Salvo, 2006, p. 106)—such epiphanies allow us to examine more carefully alternatives to the mark on the modernist wall.

The research of Bruno Latour and Steve Woolgar, carried out from 1975 to 1977 at the Salk Institute for Biological Sciences, revealed that science, although systematized in method, is informed by contingency and temporality as individual researchers authentically pursue their goals. The laboratory activities of the men behind the curtain, it appeared, were far from objective. As Latour and Woolgar discovered, scientific findings were constructed and constituted by microsocial phenomena within an agonistic, deeply human atmosphere. Personal investments made within the laboratory—the findings—were reified by writing in published studies that enhanced the credibility of the scientists who had constructed meaning within deeply contexualized circumstances. Ironically, as Latour and Woolgar demonstrate, the eventual construction of a finding masks the very circumstances (resplendent with the chance, dispersal, and indeterminacy associated with human agents) under which the research was undertaken 1979, pp. 236–244). The presence of contingency had been identified. The immutable mark on the wall was, after all, a moving snail. As Glenn J. Broadhead and Richard C. Freed (1986) found in their study of proposal writers in a business setting, the writing process may have been as linear as the scientific process, but the working conditions significantly influenced that process. Contingency and ephemerality were loose in the land.

As part of such new attitudes toward science, educational measurement also became an object of scrutiny. Questions arose regarding the assumed objectivity of educational-measurement researchers, and the realities underpinning their work were described. In 1962 Banesh Hoffman published *The Tyranny of Testing*, a fierce attack on the testing practices of ETS that paid special attention to the corporation's measurement of composition ability as a series of "fractionalized attributes" embodied in multiple-choice questions (p. 122). The attack culminated in a 1980 report sponsored by Ralph Nader, in Allan Narin's *The Reign of ETS: The Corporation that Makes Up Minds*. Legislation by New York State requiring ETS to publish its old tests, an act that forced public disclosure of test content, established the power of radical questioning.

Responsive to the ever-increasing demands for accountability during this period, Egon G. Guba and Yvonna S. Lincoln (1989) published *Fourth Generation Evaluation*. Querying extensively the traditions of educational measurement, description, and judgment, the authors described the flaws of evaluation as "a tendency toward managerialism, a failure to accommodate value pluralism, and overcommittment to the scientific paradigm of inquiry" (pp. 32–32). Within an orientation similar to that employed by Latour and Woolgar, Guba and Lincoln offer a distinctly constructivist vision of assessment—that is, an alternative model—asserting that realities are ontological social constructions of the mind and that there are always epistemological interactions between human observers and that which they observe. Methodologically, therefore, a hermeneutic inquiry is required to replace the controlling and manipulative approach too often associated with the term "experimental." Replacement is accomplished by a process that takes full advantage of the contingencies in play so that researchers may account, authentically, to all who rely on assessment as a vehicle supplying credibility to educational efforts.

It is no surprise that the interpretations of the act of technical writing emerging during this period are congruent with the emphasis on contextuality evident in the work of Latour and Woolgar in natural science and Guba and Lincoln in educational measurement. In their 1983 volume, *New Essays in Technical and Scientific Communication: Research, Theory, and Practice*, Paul V. Anderson, R. John Brockmann, and Carolyn R. Miller note that the institutional context of the teaching of technical writing had shifted. Pedagogically, the semi-interested participants identified by Donald Cunningham had disappeared. "More and better scholars with more and better preparation are entering the field," the editors wrote, "and many are now willing to put serious intellectual energy into the study of technical and scientific communication" (p. 9). Conceptually, the elementalist definition of technical writing had also been rejected. Because scientific discourse is inconsequential if assumed to correspond unambiguously to the factual world, informed research in the field was needed, the result of which would, hopefully, lead to a rejection of the "adages and commonsense assumptions" that constituted previous investigations (pp. 9–10).

The rejection of modernism—and, implicitly, rejection of the construct of technical writing as a transparent reflection of the scientific method—is evident in David Dobrin's contribution to the 1983 volume. Dobrin begins with a statement that rejects the sufficiency of experience. "There is no reason," he writes, "to believe that [those who define technical writing] have experiences that are complete, nor to believe that we can get at their experience in its totality with a few well-chosen words" (p. 229). Dobrin proposes a monadist view of language, one that acknowledges that language is attenuated, colored, and subjective. We must banish, he argues, the Cartesian rationalism, the Great Society of Western epistemology; we must accept the true nature (indeterminate in origin, dispersed in outcome) of the huddled human masses who use language. Acknowledging and accommodating the power of technology to permeate daily life will show us, Dobrin believes, that technical writing is nothing more than "a residue of technological management" (p. 243). The pride that accompanied the modernist project—the identification of technical writing with science—is now reduced to ash. Gone, Dobrin admits, are the definitions that made us "feel very comfortable with technical writing," leaving us "without much equilibrium and with an uncertain future" (p. 248). Gone are the modernist certainties, including a sense of connection that accompanied the link between the project of science and its connection with technical writing. Instead of being part of scientific progress, technical writers were, according to Dobrin, little more than stenographers of technological residue. It was as dark and demoralizing a vision as any offered by Jacques Derrida (1993). The only remedy was to abandon assumptions of closure associated with the monolith of science and to seek a new future.

The project of postmodernism thus settled into place for technical communication. The aim of the organization founded by Cunningham and his colleagues was to serve the modernist project at hand: technology. The postmodern project of technical writing, launched by Dobrin, was to question radically the role of the technical writer and to examine remorselessly the contents of documents. In the brave, new world of postmodernism, it was as if nothing at all was known before.

Emblematic of this movement—and of the postmodern view of technical communication—was the 1982 study by Lee Odell and Dixie Goswami, "Writing in a Non-Academic Setting." The researchers began, startlingly, by acknowledging that "we know relatively little about the nature and function" of non-published, day-to-day work—a statement that would have been unimaginable to T. A. Rickard's readers and their belief in the immutable power of engineering. If, as Odell and Goswami claimed, we have limited information about the variety of tasks that adult writers perform in work settings and even less information about the types of stylistic and substantive choices that writers make, there was nothing to do but to abandon assumptions and seek a new future by exploring the currents of change.

In place of a normed national sample of thousands located within a wide geographic sample were 11 participants in a government social-services agency—just the right number for a study determined to describe the day-to-day actions of day-to-day writers. Here was more than a quantitative-versus-qualitative duality. Here was a radical shift away from the identification of transcendence (science and scientific writing) to the description of immanence (social service case workers and their memos). In place of the elementalist concepts technical writing embodied in the DSST test of Technical Writing were nuanced questions designed to match the complexity of writing itself. "When writers make judgments about style," Odell and Goswami wondered, "does their sense of what is acceptable vary according to the type of writing (and implicitly, the audience and purpose) they are examining?" (p. 203). Instead of employing multiple-choice questions that would result in underrepresentation of the construct of writing and the circumstances that shaped it, the researchers interviewed the participants to gather information about the types and significance of their writing activities.

Later in the process, the researchers conducted document-based interviews about the kinds of choices the writers made and the reasons for those choices. Under conditions that were not controlled, samples of their most typical writing were collected from each participant. These documents were then rewritten by the researchers into three versions—one revised to have an active agent for all sentences, one revised to utilize the passive voice, and one in which the verbs in the active voice had been nominalized—and the participants were asked which version was most, less, and least acceptable. As part of the process of determining participant justification, Odell and Goswami noted that it was very unusual for the writers to justify a given choice by citing an a-rhetorical rule that could be followed in all circumstances; instead, most of participants made choices according to rhetorical reasons, choices made in response to concern for subject, speaker, and audience. Measures of syntactic variance (T-unit counts) demonstrated that even the features of the writing shifted with document type.

In place of the clear broth of the DSST in Technical Writing, here was gumbo. As Latour and Woolgar had found, documents were constructed and constituted by microsocial phenomena within an agonistic, deeply human atmosphere. The credibility of the writer, constructing meaning in a deeply contextualized circumstance, was deeply attuned to audience. Resplendent with the chance, dispersal, and indeterminacy associated with human agents, the writing processes and products varied within the social service agency from one group to another. By launching an investigative method in which the presence of the researchers was everywhere, Odell and Goswami discovered a complex, hypervigilant world of audience, self, and subject—a world that by its very existence had the potential to displace the vicissitudes identified with modernism. In 1986 Odell and Goswami edited *Writing in Nonacademic Settings*, a landmark book that ushered in a new world of investigative freedom.

ASSESSMENT IN THE TWENTY-FIRST CENTURY: FORCES IN PLAY

Good old fashioned value dualism gets the blood rushing. In the first decade of the 21st century, we have either the purposeful, modernist world of the objective test or the diverse, postmodern world of the case study. Pick one. Assent to that one and you are sane; demur, and you will be straightaway dangerous and handled with a chain. Yet, of course, it is the interplay—not the either/or choice—that is engaging and informative.

In the landscape of contemporary technical writing assessment, both modern and postmodern traditions are alive and well. As one of 37 DANTES Subject Standardized Tests (DSST), the Technical Writing test—claiming that it measures the ability of a student to organize and express ideas clearly with appropriate word choice and sentence structure—purrs along as smoothly as if Lee Odell, Dixie Goswami, Glenn J. Broadhead, and Richard C. Freed had never been born. In 2007 the DSST tests, managed by their once and present ETS owner, are used by community and 4-year colleges, as well as the Department of Defense. Annually, over 90,000 DSSTs are administered to those interested in continuing their education (Thomson Prometric, 2008.) In that the DANTES mission is "to support the off-duty, voluntary education programs of the Department of Defense (DOD) and to conduct special projects and development activities in support of education-related functions of the Department," it is equally significant to note that $606.6 million was allocated by the DOD in 2005 for voluntary education (DANTES, 2007). In that year, 49,071 military personnel took DSSTs (DANTES, 2005). Here is one clear path to truth.

In the postmodern camp, we may identify—with Odell, Goswami, Broadhead, and Freed—Jo Allen's treatment of the varied roles of the assessment in technical communication (1993). There is the special issue of *Technical Communication Quarterly*, edited by Hundleby, Hovde, and Allen (2003). Especially notable in that volume is the call by Michael Carter, Chris M. Anson, and Carolyn R. Miller (2003) to see the isolated, postsecondary technical writing service course as part of the entire curriculum that shapes a student, to imagine the course as a way to embrace the values of openness, participation, and dispersal that are inherent in language use. Here are the very values that are absent in the orientation of the DANTES Technical Writing test toward closed practice, objectified purpose, and centered discourse. Here are dark, deep, lovely woods.

Certainly, it is true that the structuralist world of testing gleams in all its modernist certainty alongside of the contextual world of assessment in all its postmodern contingency. It is also equally true, however, that the marketplace supporting each edifice is rapidly changing, and that there is currently a transformation—both economic and political in nature—in which flexible accumulation is rapidly becoming the order of the day. The postwar boom from 1945 to 1973, built on a Fordist-Keynesian base, is being displaced by shifting markets that thrive by means of flexible labor practices that are, at once, entrepreneurial in spirit and global in domain (Harvey, 1989, pp. 121–172; Frieden, 2006; Friedman, 2007). It is the need for just such flexibility that Michael J. Salvo identifies when he writes about technical communicators inventing opportunities for cultural power and authority. If we recall that 1,439,264 bachelors degrees were conferred in the United States in 2005 and that 616,273 of these degrees were in business, communication, computer and information science, engineering, engineering technologies, English, and the liberal arts (National Center for Education Statistics, 2006, Table 254)—all fields hosting courses in technical communication—then we can safely conclude that instructors of technical communication have the capacity both to instruct and assess the advanced literacy ability of 42.8% of the nation's graduates. In the technical writing service course alone, as Carter et al. (2003) note, the high-demand, multiple-section offerings have a capacity equal to the first-year composition course, a capacity to foster a proactive response that "begins with questions and responds with the analysis of data, leading to continuous self-study without the need for external motivation" (p. 105). As a profession, our portfolio for outcomes assessment is impressive. But to invent such opportunities, we must clarify our vision and capitalize on our strengths.

As causal relationships go, the profession of technical communication has escaped the great assessment wars that have so engrossed the profession of composition. The composition battles that have been fought since the College Entrance Examination Board administered its first tests in 1901 have not been part of the teaching of technical writing. Because students have already been accepted into the academy when they enroll in technical communication classes, the complex socioeconomic decisions of admission and placement are not central to the world of technical communication. Thus, while academic writing assessment has a century's worth of history, the present volume is the first book-length study designed to address the aims and methods of technical writing assessment. Growing as it did out of the Hundleby et al. (2003) *Technical Communication Quarterly* issue, the present book may be seen as congruent with the demands of organizations such as the Accreditation Board for Engineering and Technology (2006), which has begun including the assessment of educational outcomes as part of its review process.

These demands are new and have not yet fully integrated themselves into postsecondary education. Such gaps between performance and innovation are further evidenced in nonacademic settings. Heavily regulated industries such as pharmaceutical manufacturing have not yet targeted technical writing for review—although the Code of Federal Regulations (2008) clearly defines stringent requirements for control documents as part of the manufacturing record (Elliot, Haggerty, Foster, & Spak, 2008). Regardless of setting, demands for accountability are increasing. While the profession of technical writing has not yet identified a strong presence within the assessment arena, the time is coming when both academic and nonacademic accountability will become part of daily life within a flexible economy that will increasingly employ outcomes assessment methods. If there is no assembly line with a Ford automobile at the end serving to validate effectiveness and efficiency, then other methods of judgment will be sought.

Identification of the central constructs of technical communication will be significant to the profession's success in this new world. Questions about agents and agencies will be asked, and the answers given will be highly influential in determining future validity claims for our field. A compelling case for defining technical communication as an act of articulation can be presented (Thayer, 2002), but in the postmodern world, it may also be best to avoid the overarching definitions of modernism and, instead, let the agents decide for themselves (Allen, 1990). It is not so much that the attempt to define is too complex, but rather that courses and programs really do serve the needs of different constituencies (Allen, 1993, p. 366). It is not that researchers such as Allen are against concrete definitions; rather she would prefer to allow definitions to arise that are intrinsically valid for different regions, institutions, and students. Believing with the feminist-standpoint communication theorists that no two people live the same experiences (Buzzanell, 1994, pp. 353–355), Allen's refusal to define is a refusal to accept the totalitarian aspects of definition-making. Hers is a call to make our definitions relevantly visible to each other. If the profession follows such a decentralized model, then there need not be worry about who is setting the standards. The answer is that we are.

The road to establishing such a decentralized model stretches out before us. The Council for Programs in Technical and Scientific Communication advocates review rather than accreditation, but that orientation will falter in larger academic and corporate arenas if we do not move quickly in developing a material demonstration of the credibility of our practice by articulation of our constructs and agendas. The 2006 report by Secretary of Education Margaret Spellings, *A Test of Leadership: Charting the Future of U.S. Higher Education*, is ominous in its urge to assess expediently our outcomes for us.

An excellent example of articulation that is discipline-based and meaningful is found in the Council on Linkages Between Academia and Public Health Practice (2009). Since 1992 the Council has worked to develop a set of core competencies based on key variables for public health professionals. Reviewed by over 1,000 professionals, the core competencies are not to be used as a set of rules to be followed, but instead, as "a starting point for developing a modified list of competencies" that matches the needs of users. During this period, the Council has established a national forum in the areas of academic and practice linkages, workforce competencies, public health systems research, and worker recruitment and retention. Key to the cohesion of the Council is the ongoing development of the competencies and the way that specific institutional sites modify, use, and assess these competencies. Surely, similar work could be hosted by the Association of Teachers of Technical Writing (ATTW, 2007), the Council for Programs in Technical and Scientific Communication (CPTSC, 2007), and the Society for Technical Communication (STC, 2007). Indeed, at present the STC is beginning to take steps to define a body of knowledge for the profession (Dayton et al., 2007). Surely assessment programs, based on core competencies and modified to meet the needs of different stakeholders, soon must be designed in our visionary imaginations before we find them constructed, in reality, before us.

The assessment of visual communication within a computer-mediated environment is an excellent venue for innovation. Within the construct of computer-mediated visual communication, the human interaction with language, ideas, and representation is, at once, most diverse and most unified. With its history of research into the impact of technology and the need for informed usability practices, the profession shows its strongest hand in this area. It is significant that Lee Odell, who has presciently led in so many areas of assessment for so long, has brought forward *Writing in a Visual Age* (2006) with Susan M. Katz. With its emphasis on the look of document, the design of the textbook itself—a playful collage of text, image, and color—prompts readers to find and promote their own unique voices. Implicitly, such work urges readers to reject their own transparency as vehicles of technology and adopt new rhetorical roles as agents of change.

Ultimately, assessment can only benefit from the study of these roles because so much remains to be known about the forces—from aesthetic to axiological— that impact computer-aided visual communication, an integral part of technical communication in the 21st century. While such studies may have proven daunting when Guba and Lincoln launched their case for a new generation of contingency-based assessment, new conceptualizations of the process of gathering validity evidence are now emerging (Kane, 2006). With validity now viewed as a series of integrated judgments (Messick, 1989; AERA, APA, & NCME, 1999) and as a process similar to theory development (Kane, 2001), a new world of evaluation is before us.

CONCLUSION: PATHS TAKEN

It is, after all, time to pick. Modernism or postmodernism? As a radical act, pick the best of both. Consider the following scenario: driving a Honda Accord Hybrid from a suburban home, parking the car at City College, and walking to the second floor of an office in a building designed by Michael Graves. The day is full, divided into two parts. The morning is set aside to score e-portfolios in a collaborative assessment project designed to pilot-test a new rubric intended to help instructors rate the performance of students completing a graduate technical writing seminar. The work is inspired by the notions of community offered by Alan C. Purves in *The Web of Text and the Web of God: An Essay on the Third Information Transformation* (1998). Collaborators join in from across the country. An afternoon meeting with two colleagues—one in computer-based graphic design and one in nutrition—focuses on the development of new ways of visualizing dietary approaches designed to stop hypertension. This work is informed by Lee. E. Brasseur's *Visualizing Technical Information: A Cultural Critique* (2003). If the culturally based usability studies are effective, the new design will be posted on the Web site of the city's Department of Health and Human Services.

A kinda modernist morning, a sorta postmodern afternoon—it would, after all, be a good day.

REFERENCES

Accreditation Board for Engineering and Technology (ABET). (2006). *Accreditation policy and procedure manual*. Baltimore, MD: ABET.

Allen, J. (1990). The case against defining technical writing. *Journal of Business and Technical Communication, 4*, 68–77.

Allen, J. (1993). The role(s) of assessment in technical communication: A review of the literature. *Technical Communication Quarterly, 2*, 365–388.

American Council on Education (ACE). (1987). *Guide to educational credit by examination* (2nd ed.). New York: Macmillan. (Princeton: ETS Archives)

American Educational Research Association (AERA), American Psychological Association (APA), & National Council on Measurement in Education (NCME). (1999). *Standards for educational and psychological testing*. Washington, DC: American Psychological Association.

Anderson, P. V., Brockmann, R. J., & Miller, C. R. (Eds.). (1983). *New essays in technical and scientific communication: Research, theory, practice*. Amityville, NY: Baywood.

Andrews, D., & Blickle, M. (1982/1978). *Technical writing: Principles and forms* (2nd ed.). New York: Macmillan. (Original work published 1978)

Association of Teachers of Technical Writing (ATTW). Home Page. (2007). Retrieved November 4, 2007, from http://cms.english.ttu.edu/attw

Brasseur, L. (2003). *Visualizing technical information: A cultural critique*. Amityville, NY: Baywood.

Brigham, C. (1923). *A study of American intelligence*. Princeton, NJ: Princeton University Press.

Broadhead, G., & Freed, R. (1986). *The variables of composition: Process and product in a business setting*. Carbondale and Edwardsville, IL: Southern Illinois University Press.

Brusaw, C., Alred, G., & Oliu, W. (2003). *Handbook of technical writing* (7th ed.). New York: St. Martin's. (Original work published 1976)

Buzzanell, P. M. (1994). Gaining a voice: Feminist organizational communication theorizing. *Management Communication Quarterly, 7*, 339–383.

Carter, M., Anson, C. M., & Miller, C. R. (2003). Assessing technical writing in instructional contexts: Using outcomes-based assessment for programmatic thinking. *Technical Communication Quarterly, 21*, 101–104.

Chauncey, H. (1958). *Educational testing service annual report, 1957–1958*. Princeton, NJ: ETS.

Chauncey Group. (2002). Fact sheet. Princeton, NJ: The Chauncey Group. Retrieved November 4, 2007, from http://www.getcollegecredit.com/06colleges_d.html

Code of Federal Regulations. (2008). 21CFR 211 Subpart J. Washington, DC: Government Printing Office. Retrieved November 4, 2007, from http://www.accessdata.fda.gov/scripts/cdrh/cfdocs/cfcfr/CFRSearch.cfm?CFRPart=211&showFR=1&subpartNode=21:4.0.1.1.10.10

Connors, R. (1982). The rise of technical writing in America. *Journal of Technical Writing and Communication, 12*, 329–352.

Council for Programs in Technical and Scientific Communication (CPTSC). Home Page. (2007). Retrieved November 4, 2007, from http://www.cptsc.org/

Council on Linkages Between Academia and Public Health Practice. (2009). *Core competencies for public health professionals*. Retrieved May 12, 2009, from http://www.phf.org/link/competencies.htm

Cunningham, D. (2004). The founding of ATTW and its journal. *Technical Communication Quarterly, 13*, 121–130.

Dayton, D., Davis, M., Harner, S., Hart, H., Mueller, P., & Wagner, E. (2007, September). *Defining a body of knowledge*. Houston, TX: Society for Technical Communication Academic-Industry Leaders Summit, University of Houston.

Defense Activity for Non-Traditional Educational Support (DANTES). (2005). *Voluntary education fact sheet, fy 2005*. Retrieved November 4, 2007, from http://www.dantes.doded.mil/Dantes_web/library/docs/voledfacts/FactSheetArchive.pdf

Defense Activity for Non-Traditional Educational Support (DANTES). (2007). *Mission*. Retrieved November 4, 2007, from http://www.dantes.doded.mil/dantes_web/danteshome.asp?Flag=True

Derrida, J. (1993). *Aporias* (T. Dutoit, Trans.) Stanford, CA: Stanford University Press.

Dobrin, D. (1983). What's technical about technical writing? In P. V. Anderson, R. J. Brockmann, & C. R. Miller (Eds.), *New essays in technical and scientific communication: Research, theory, practice* (pp. 227–250). Amityville, NY: Baywood.

Eagleton, T. (1996). *Literary theory: An introduction* (2nd ed.). Minneapolis, MN: University of Minnesota Press.

Educational Testing Service (ETS). (1985). Fact sheet/study guide, technical writing SE-820. Princeton, NJ: Educational Testing Service. (Princeton: ETS Archives)

Educational Testing Service (ETS). (1991). *Fact sheet/study guide, technical writing SE-820.* Princeton, NJ: Educational Testing Service. (Princeton: ETS Archives)

Educational Testing Service (ETS). (1994). *Fact sheet/study guide, technical writing SF-820.* Princeton, NJ: Educational Testing Service. (Princeton: ETS Archives)

Educational Testing Service (ETS). (1998). *Fact sheet/study guide, technical writing SF, SG-820.* Princeton, NJ: Educational Testing Service. (Princeton: ETS Archives)

Eliot, C. W. (1930). President Eliot's inaugural address. In Samuel Eliot Morison (Ed.), *The development of Harvard University since the inauguration of president Eliot, 1869–1929* (pp. lvix–lxxxviii). Cambridge, MA: Harvard University Press. (Original work published 1869)

Elliot, N., Haggerty, B., Foster, M., & Spak, G. (2008). Asynchronous training in pharmaceutical manufacturing: A model for university and industrial collaboration. *International Journal on e-learning, 7,* 67–85.

Estrin, H. (Ed.). (1963). *Technical and professional writing: A practical anthology.* New York: Harcourt.

Frieden, J. (2006). *Global capitalism: Its fall and rise in the twentieth century.* New York: Norton.

Friedman, T. L. (2007). *The world is flat: A brief history of the twenty-first century.* New York: Farr, Straus and Giroux.

Guba, E., & Lincoln, Y. (1989). *Fourth generation evaluation.* Newbury Park, CA: Sage.

Harris, J. S. (1978). On expanding the definition of technical writing. *Journal of Technical Writing and Communication, 8,* 133–138.

Harvey, D. (1989). *The condition of postmodernity: An enquiry into the origins of cultural change.* Cambridge & Oxford, UK: Blackwell.

Hassan, I. (1985) The culture of postmodernism. *Theory, Culture, and Society, 2,* 119–131.

Hegel, G. W. F. (1975). *Lectures on the philosophy of world history* (H. B. Nisbet, Trans.). New York: Cambridge University Press. (Original work published 1837)

Hoffmann, B. (1962). *The tyranny of testing.* New York: Collier.

Houp, K. W., Pearsall, T. E., Tebeaux, E., & Dragga, S. (2005). *Reporting technical information* (11th Ed.). New York: Oxford University Press. (Original work published 1968)

Hundleby, M. N., Hovde, M., & Allen, J. (Eds.). (2003). Assessment in technical communication [Special Issue]. *Technical Communication Quarterly, 12*(1).

Kane, M. T. (2001). Current concerns in validity theory. *Journal of Educational Measurement, 38,* 319–342.

Kane, M. T. (2006). Validation. In R. L. Brennan (Ed.), *Educational measurement* (4th ed., pp. 17–64). Westport, CT: American Council on Education and Praeger.

Lannon, J. M. (1999). *Technical writing* (7th ed.). New York: Scott, Foresman. (Original work published 1982)

Latour, B., & Woolgar, S. (1986). *Laboratory life: The construction of scientific facts.* Princeton, NJ: Princeton University Press. (Original work published 1979)

Messick, S. (1989). Validity. In R. L. Linn (Ed.), *Educational measurement* (3rd ed., pp. 13–103). New York: American Council on Education and Macmillan.

Mills, G. H., & Walter, J. (1986). *Technical writing* (5th ed.). New York: Holt. (Original work published 1954)

Nairn, A. (1980). *The reign of ETS: The corporation that makes up minds.* Washington, DC: Nader.

National Center for Education Statistics. (2006). Table 254: Bachelor's degrees conferred by degree-granting institutions, by discipline division: Selected years, 1970–71 through 2004–05. *Digest of Education Statistics.* Retrieved November 4, 2007, from http://nces.ed.gov/programs/digest/d06/tables/dt06_254.asp?referrer=list

Odell, L., & Goswami, D. (1982). Writing in a non-academic setting. *Research in the Teaching of English, 16*, 201–223.

Odell, L., & Goswami, D. (Eds.). (1986). *Writing in nonacademic settings.* New York: Guilford.

Odell, L., & Katz, S. M. (2006). *Writing in a visual age.* Boston, MA: Bedford/ St. Martin's.

Pound, E. (1934). *Make it new: Essays.* London: Faber and Faber.

Purves, A. C. (1998). *The web of text and the web of God: An essay on the third information transformation.* New York: Guilford.

Rickard, T. A. (1931). *Technical writing* (3rd ed.). New York: McGraw. (Original work published 1920)

Salvo, M. J. (2006). [Review of the book *Strategies for professional status*]. *Technical Communication Quarterly, 15*, 103–108.

Sandman, P. M., Klompus, C. S., & Yarrison, B. G. (1985). *Scientific and technical writing.* New York: Holt.

Schwager, D., Gall, L., & Thomas, S. (1995). *Technical summary: Technical writing.* SR 95-43. Princeton, NJ: Educational Testing Service. (Princeton: ETS Archives)

Society for Technical Communication (STC). Home Page. (2007). Retrieved November 4, 2007, from http://www.stc.org/

Spellings, M. (2006). *A test of leadership: Charting the future of U.S. higher education.* Washington, DC: U.S. Department of Education. Retrieved November 4, 2007, from http://www.ed.gov/about/bdscomm/list/hiedfuture/reports/final-report.pdf

Thayer, A. (2002). Defining technical communication: Is it a goal or a Sisyphean task? *Orange: An Online Journal of Technical Communication and Information Design.* Retrieved November 4, 2007, from
http://orange.eserver.org/issues/3-2/thayer.html

Thomson Prometric. (2007). *Fact sheet, technical writing.* Retrieved May 12, 2009, from http://www.getcollegecredit.com/downloads/factsheets/Technical%20Writing.pdf

Thomson Prometric. (2008). *DSST Candidate information bulletin.* Retrieved May 12, 2009, from http://www.getcollegecredit.com/downloads/DSST_CIB_20090109.pdf

Todd, J. (2003). Teaching the history of technical communication: A lesson with Franklin and Hoover. *Journal of Technical Writing and Communication, 33*, 65–81.

Walter, J. A. (1977). Technical writing: Species or genus? *Technical Communication, 21*, 6–8.

Weisman, H. M. (1992). *Basic technical writing* (6th ed.). New York: Prentice Hall. (Original work published 1985)

Yerkes, R. M. (1921). *Psychological examining in the United States Army*. Memoirs of the National Academy of Sciences: Vol. 15. Washington, DC: Government Printing Office.

Assessment in the Widest View

Mapping Institutional Values and the Technical Communication Curriculum: A Strategy for Grounding Assessment

Jo Allen, Widener University

As we explore various elements of technical communication programs and assess student learning outcomes within those programs, one of the most difficult aspects may be trying to ground our curriculum in an appropriate context suitable for assessment. Without such grounding, attempts to assess learning may float among various competing contexts: other curricula at the institution, other technical communication curricula throughout the region or nation, or contexts provided in feedback from alumni or employers who describe their expectations for knowledge and ability from our graduates.

While any of these contexts may be valuable for at least the beginning stages of determining appropriate assessment strategies, I suspect that all program directors and faculty members have felt some frustration when trying to situate their programs in the context of, say, (other) nationally known programs. The disparity in resources, history, faculty strengths, institutional type, and other characteristics of these programs may well doom any attempt of a comparison to failure. For that and other reasons, most assessment experts caution against one-size-fits-all assessment (see, for instance, Anderson, 2004; Astin, 1993; Huba & Freed, 2000); in fact, the very foundation of any assessment should be directed toward a program's values, and if the values belong to someone else's program, the resulting disconnect is not only predictable, but potentially even paralyzing.

Thus, looking at other programs may be useful in trying to design programs or trying to assess particular elements of programs, but they are rarely useful in designing a full assessment strategy. Instead, one of the most illuminating discussions may well be the intersections between the technical communication curriculum and the institution's own expressed values. Such a context—program within institution—may seem obvious, but from my years of experience in assessment and more recently as a provost, few assessment plans or strategies fully take on the institution's core values as a backdrop for their work. As a good example, many institutions that promote research as a defining aspect of their mission extend that sense only into expectations of faculty productivity, not into the individual curricula in any pervasive way, and especially not at the undergraduate level. It makes all kinds of sense that a technical communication program in a research intensive or extensive institution would reflect that value in its curriculum and thus, in its assessment; yet, only a few do so.

Often found in mission statements or statements of core values, educational characteristics, and even general education goals, institutional values reflect the defining characteristics of a particular institution's approach to education. As such, they are highly useful in our conversations and directions for assessment, especially in assessing connections between programs and the larger institutional context. In fact, these values can and should serve as a guide or template for the kinds of qualities that faculty should consider when developing and assessing their curriculum.

At this point, we need to distinguish between an institution's mission and its goals or core values, which should most certainly be linked but are not interchangeable. At its simplest, a mission reflects what the university is (research, doctoral, liberal arts), but just as importantly, what it does (leads, provides, engages, serves, promotes); institutional values suggest its desired outcomes (civic leaders, global learners, citizens of character, industry leaders). And it is in the curriculum and co-curriculum where the two should meet, with those elements being the repository of evidence of delivering on the values. While accrediting bodies (both regional and disciplinary) require mission statements, they do not necessarily require statements of institutional values. And while they require demonstrable links between the mission and any given program, they may not require links between the institutional values (if they exist) and the program. Moreover, they typically ask for description of links, rather than assessment. So, strengthening that connection through both description and assessment may be overlooked as a particularly valuable means of situating a program in its institutional culture, while also grounding the program's assessment strategy. (For more on the relationship between missions, goals, values, and other distinguishing features of an institution, see Nichols and Nichols, 2005).

In this chapter, using a series of templates and rubrics, I intend to demonstrate ways that an institution's values might shape the assessment of a technical communication program's curriculum and co-curriculum. In making connections between institutional values and the curriculum, I hope to argue for a critical review that will highlight program strengths and points of distinction, as well as weaknesses and points for improvement. In such a review, opportunities for keeping programs dynamic, responsive, and relevant are significant and help clarify the elements of a robust educational experience that is attractive to faculty, students, administrators, alumni, donors, employers, parents, and other stakeholders.

AN EXPLORATION OF "INSTITUTIONAL VALUES," INCLUDING THE POTENTIAL FOR PROGRAM ASSESSMENT

Conversations about "institutional values" might refer to any number of influences and sources with varying degrees of formality, such as the examples that follow:

1. *the heritage* that inspires the institution's traditions (e.g., historically black colleges and universities; the founders' motivations and vision for the institution, women's colleges);
2. *the traditions* that shape the institution's spirit, reputation, and culture (e.g., a highly rated athletic program, a highly charged atmosphere of political activism, a "party school");
3. *the founding or subsequent mission* or charge of the institution (e.g., a military college that educates military officers, a religious college that educates clergy or religious employees, a trade school that educates a particular segment of workers);
4. *the founding or subsequent curriculum* that defines the character of the institution (e.g., fine arts, agriculture, engineering and technology, liberal arts).

Such contexts that may shape institutional values are typically embedded in and produce all sorts of manifestations throughout the college community, appearing as evidence in the curriculum, residence halls (or lack thereof), co-curricular programming, student activities, the population base, student demographics or profiles, and so on. At a curricular level, however, the institution's values, if articulated, may well shape the distinctive qualities of the education offered. Therefore, the values might reflect some singular quality or, just as likely, a combination of defining characteristics such as

- an emphasis on the arts;
- an emphasis on bench or applied research;
- an emphasis on civic engagement, community service, volunteerism, or nonprofit organizations;
- an emphasis on communication (writing and speaking across the curriculum);
- an emphasis on the needs of a particular group of students (adult learners, women, artists);

- an emphasis on technology/technical applications;
- an emphasis on teaching/learning methodologies: active learning, experiential learning, distance learning, global learning;

plus any number of other special focuses.

In articulating their values, the faculty and administration (and Board of Trustees) may create the kind of statement that reflects student characteristics (the kinds of student we educate: commuters, adult learners, first-generation college students, women), disciplinary specializations (in these disciplines: technology, liberal arts, engineering, religious studies), noting any special contexts (evening and weekend classes, online education, experiential learning, applied research), with an eye toward specific outcomes (to create lifelong learners engaged in community service, to create engineers with global perspectives, to graduate technical experts with artistic sensibilities). As these elements coalesce, the institutional values emerge as a clear indication of the context for any given curriculum. Using those values, therefore, to move from the institutional to the programmatic level of assessment is a key strategy for connecting a technical communication program to its institutional roots.

MAKING THE CONNECTION: INSTITUTIONAL VALUES AND TECHNICAL COMMUNICATION PROGRAMS

For the remainder of this discussion, several elements can demonstrate a set of core or institutional values that define the kind of education a college or university might offer. Building on the institution's core values, we can create templates and rubrics (discussed below) that highlight the characteristics of the institution's technical communication program within that particular institution's context, noting the degree to which the program addresses (and eventually assesses) those values. The following template, Table 1, provides several examples of possible institutional values along the y-axis. It is not intended to suggest that a single institution might hold all of these values, although such may be the case. What is intended to be of value here is the demonstration of connections (or maps) linking the institution's values and how they manifest themselves in technical communication program values, along with where lessons about those values appear in the technical communication curriculum.

Ironically, some institutions have programs whose values never reflect the institution's values or, when they do, cannot point to a place in the curriculum where those values are actually explored and taught. It is especially valuable at this point, therefore, to note where the value is implicitly versus explicitly taught. For instance, the difference between a course in writing and a course in which writing is required and thus, one means of assessing students' content knowledge, may well be two different animals. Assessing what students produce in a course, separate from the context of what they are taught, can be sticky when we try to direct our questions about learning to specific points in the curriculum. The issue of primary versus secondary emphasis in content, accountability, and even faculty expertise is important here; for if students are expected to demonstrate, say, fluency in proposal writing but never take a course or have extended instruction in proposal writing—instead, have only courses that require proposals to be written during the semester—that may signal the kind of disconnect between expectations and actual instruction that is especially illuminating in the assessment context. What faculty decide to do to remedy that disconnect is one of the greatest demonstrations of the role assessment plays in promoting faculty ownership of program improvements.

MAKING IT PERSONAL:
THE TECHNICAL COMMUNICATION IMPRIMATUR

Next, the template calls for a general description of the objectives and potential outcomes of each institutional value as it manifests itself in the technical communication curriculum—in short, what will this institutional value look like in the context of our technical communication program? While a separate format such as that laid out in Figure 1, next section, can be constructed to outline the objectives/outcomes/measures and other particulars of the actual assessment plan, we can at least gather a sense of those expectations from this original template. Next, and as a critical reminder to us all, it is imperative that we plot sites of learning into this template. While we may articulate our expectations of program outcomes in measurable ways, we must also document where in the curriculum we are actually teaching the content, processes, and products that will surface those outcomes. In fact, some of the most valuable realizations in assessment work have come when faculty have discovered that some of their grandest expectations about student learning do not actually have a site in the curriculum; the assumption that somebody else was teaching a concept or even that everybody else was teaching it is finally jettisoned in the realization that, actually, nobody is teaching that content. Such gaps are further evidence of the ways that assessment can improve programming.

Next, the column for "co-curricular contact" is designed to acknowledge ways that student learning is promoted outside the classroom. Many institutions' student services, student affairs, or other co-curricular divisions are directly engaged in promoting programming to enhance learning opportunities that clearly adopt the institution's values. Residential learning communities, for instance, are one of the most prominent features of the co-curricular emphasis on learning outcomes in support of the curriculum and the institution's values. Other opportunities for student engagement and service are also good examples of such sites of learning and, thus, deserve incorporation to both the plan for learning and the assessment.

Table 1. Template for Connecting Institutional Values to Individual Programs

Institutional value	General outcomes	T.C. program outcomes	Curricular contact	Co-curricular contact	Degree of proficiency
Communication skills [written]	Can explain common disciplinary issues, approaches, themes, methods, etc. to expert and lay audiences; adapt the info. to the primary genres of the major discipline, make editing decisions that strengthen a document; correct major errors in standard English; identify places in text where graphic elements would enhance the clarity or impact of the message.	Can translate any subject to any group of readers, ranging from expert, to managers, to special interest audiences, to lay readers; use any document format; make full stylistic, rhetorical, design, and other document edits; identify and correct grammar, spelling, and mechanical errors; incorporate graphic elements using current technology to enhance clarity, impact, etc.	ALL courses, TC 101, 102, ETC.	Internship, service learning and community work/documentation	Mastery
Communication skills [oral]	Can explain common disciplinary issues, approaches, themes, methods, etc. in formal presentations; adapt information for informal presentations, including small to large groups that may or may not have any decision-making authority; adapt information for various audiences and for exploratory, explanatory, or	Can explain any subject to any group of listeners, ranging from expert to managers to special-interest audiences to lay readers; readily ascertain the appropriate level of formality in the group and adapt information accordingly; facilitate group discussions based on information presented; manage the details and specifics, as well	ENG 101, 102 TC 211-222 TC 423 TC 599	Presentations for internships; community work, student organizations (STC Student Chapter)	Mastery

persuasive presentations; present ideas in a well-organized, articulate fashion; incorporate appropriate examples, details, graphics, and other elements that help clarify the oral presentation; handle basic Question & Answer sessions.

as the overall scheme and impact of the topic at hand; accommodate hostile audiences; provide appropriately documented and referenced works to establish credibility and authority; easily use graphic and presentation technology to enliven and clarify information; handle Question & Answer sessions with felicity.

Regional Understandings of Disciplinary Impact	Can explain the relevance of their discipline on the economic, social, political, historical, environmental, cultural or other critical venues of the region; articulate current events that have a direct impact on the practice of their discipline in the region and, conversely, articulate their discipline's current impact on at least one current issue in the region.	Can explain the relevance of technical communication on the economic, social, political, historical, environmental, cultural, or other critical venues of the region; articulate current events that have a direct impact on the practice of technical communication in the region, and, conversely, articulate technical communication's impact on at least one current issue in the region.	TC 399 TC 499	STC student chapter; community service; internships/cooperative education; other student organizations; guest speakers in classes; field trips to area businesses; interactions with alumni; career fairs; information exchange fairs

Mastery

Table 1. (Cont'd.)

Institutional value	General outcomes	T.C. program outcomes	Curricular contact	Co-curricular contact	Degree of proficiency
Global Understandings of Disciplinary Impact	Can explain the relevance of their discipline on the economic, social, political, historical, environmental, cultural or other critical venues of the world; articulate current events that have a direct impact on the practice of their discipline in various countries and conversely articulate their discipline's current impact on at least one current issue in a specific area of the world; document the source and authority for that understanding and articulate at least some of the impact for the discipline and for the U.S. and its relations with that country or region.	Can explain the relevance of technical communication on the economic, social, political, historical, environmental, cultural, or other critical venues of the world; articulate current events that would have a direct impact on the practice of tech. communication in that region, and, conversely, articulate technical communication's impact on at least one current issue in the region; document the source and the authority for that understanding and articulate at least some of the impact for the discipline and for the U.S. and its relations with that country or region.	TC 299 TC 499	Study Abroad; STC student chapter; community service, internships/cooperative education; other student organizations; guest speakers in classes/interactions with alumni	Mastery of all areas for at least one international arena

Technology	Can use the common technological tools of the discipline, understanding the appropriate situations in which to use each tool and the best means of application; articulate the reasons for using the tool as well as reasons not to use the tool.	Can use a broad array of technology for various tasks in the technical communication arena, including word processing, graphics, Web-creation, citation, spreadsheets, project scheduling, presentation, and other software packages; research topics in technical communication through standard and Web-based search processes.	ALL TC courses	Community service/ service learning as needed with student organizations.	Mastery in word processing; high proficiency in graphics & Web-creation; moderate proficiency in project scheduling and spreadsheets unless project/ career choices warrant greater proficiency.
Entrepreneurial Thinking	Can articulate clear applications of disciplinary contributions to frontier and otherwise unexplored areas of development; explain the needed resources for the development, along with the elements of attractiveness and potential downfalls or complications.	Can articulate clear applications of technical communication to frontier and otherwise unexplored areas of development from the sciences, business, and/or technology arenas; see the elements of viability and danger for those ideas; identify particular instances of people and places that might benefit from the idea; describe partners, or at least the characteristics of partners, likely to be interested in funding the idea.	TC 399 TC 499	Internships/cooperative education; community-service partners, service-learning partners; collaborative learning opportunities throughout the curriculum and co-curriculum.	Moderate to high capacity for recognizing sound entrepreneurial thinking; some moderate to high degree of proficiency in crafting at least one substantive entrepreneurial idea.

SAMPLE ASSESSMENT PATTERN

Institutional Value: Communication skills [Written]

Program: Technical and Professional Communication

Objective 1: Students can translate any subject to any group of readers, ranging from expert to managers, to special interest audiences, to lay readers.

Outcomes: 1. Students make, and can defend, good choices about the amount of detail an audience expects (on the advanced side) or can process (on the general or lay side) of any topic.

2. Students make, and can defend, good choices for appropriate vocabulary that is suitable for the comprehension level of the audience.

3. Students make, and can defend, good choices for organization of information that is suitable for the comprehension level of the audience.

Objective 2: Students can make full stylistic, rhetorical, design, and other document edits.

Outcomes: 1. Students can articulate the rationale for their edits as they move between versions of documents.

2. Students can make well-reasoned edits in tone, organization, evidence, persuasive strategy, format and other elements as needed.

3. Students can accommodate the intricacies of these edits according to the document's purpose, audience, and urgency and can articulate the rationale for their decisions.

4. Students can move fluidly between editing and proofreading to address any modifications that need to be made.

Measures: 1. A portfolio of student work is submitted at the end of the junior and senior years.
2. An extensive final project is part of the senior capstone experience. A defense of that project before the faculty assures that the student can articulate the rationale for choices.

Criteria for Evaluation

Portfolio:

- The selection of work demonstrates developing sophistication of the writing in terms of fluency with style, rhetorical strategies, design, and other features.
- The selection of work demonstrates fluency with genres.
- The selection of work demonstrates thorough understanding of how to bring together the content, context, audience, purpose, and other features to create useful documents.
- The selection of work demonstrates a thorough understanding of the conventions of standard English or the equivalent language.

Final Project:

- The final project offers an extended example (at least 10 pages or the equivalent in a web-based document) of the student's capabilities for managing a sustained work.
- The final project demonstrates the student's ability to
  - clarify a subject and its context
  - identify and accommodate the relevant characteristic features of the audience
  - select the appropriate genre for the project, incorporating all its standard features
  - select the appropriate design features that support the genre, topic, audience, and other critical features
  - defend the selection of critical stylistic, rhetorical, design, and other features of the document

Figure 1. Articulation of objectives, outcomes, and measures in the context of institutional values.

At the far right of the template is a notation of the expectation of competence for each value. In some cases, for some curricula, we might reasonably expect higher levels of competence than in others. We should surely, for instance, expect more critical elements of mastery of written communication skills in a degree program in technical communication than we might expect in other nonwriting majors; similarly, we might be a bit more forgiving in some of our expectations of students' other skills and learnings, determining that "moderate" or even "beginning" knowledge is acceptable—in other words, that we are looking, in some cases, for awareness, rather than mastery.

ARTICULATING THE CONNECTION IN A WELL-DEVELOPED ASSESSMENT PLAN

Following the template with its links between institutional values and their manifestation in the technical communication's values and curriculum is the actual assessment plan, shown here in Figure 1 as an outline of objectives, outcomes, measures, instruments, and other particulars. At this point, there is nothing new, as dozens of experts in assessment have drawn any number of formats for articulating objectives and outcomes. What is new, for some, may be that we are now documenting the specifics of our expectations as they reflect those institutional values of learning: What are the grand visions—the objectives—for technical communication students' knowledge and performance abilities in the context of our institutional values? How do they manifest themselves as particular outcomes? What kind of artifact might we require as the evidence (either singular or collective) of that learning? On what criteria might we establish our evaluation of the students' work?

What this figure ultimately shows is a schema for moving from the articulation of a desired learning objective to the specific outcome(s), the measures providing evidence of learning, the artifact and criteria for evaluation, and elements of accountability (timelines and reporting responsibilities), simply demonstrated as the familiar assessment cycle in Figure 2.

The seamlessness of the connection as demonstrated here, of course, should not be misconstrued to trivialize the work of assessment or the students' learning processes. Both processes are highly complex and depend on any number of variables. For assessment, matters of faculty buy-in and expertise, resources and support, and education in assessment affect the work. For determining the impact of the curriculum on student learning processes, measurement is equally complex, focusing on issues as broad and pervasive as the students' socioeconomic and educational background, all sorts of "exposures" to individual interest and learning styles, motivation, and so on. At a programmatic level, however, where both student and faculty circumstances are tempered by the larger perspective on student learning and student needs, those individual characteristics may easily,

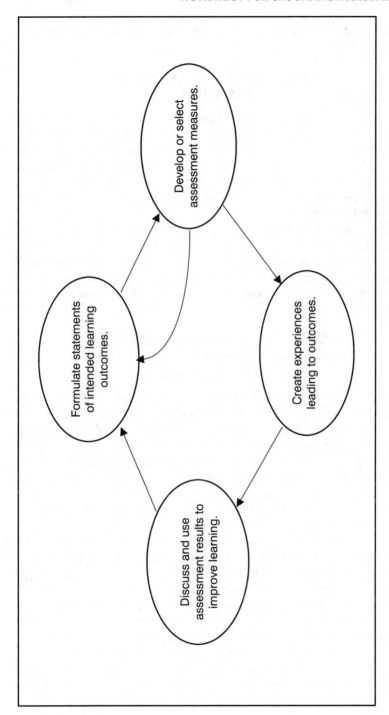

Figure 2. Assessment cycle by Mary Huba and Jann C. Freed.
(Adapted from Huba & Freed, *Learner-Centered Assessment on College Campuses:
Shifting the Focus from Teaching to Learning*)

and valuably, be morphed into a larger perspective on student abilities and program renderings.

MOVING FROM EXPECTATION TO EVALUATION

At the end of any assessment strategy should be answers to the question, "How are we doing?" In programmatic assessment, we are looking for answers that match the extension of institutional values through programmatic values into objectives and outcomes and result, ultimately, in the determination of learning. Thus, Figure 3 articulates the distinctions between the expected levels of mastery, typically ranging from "novice" or "beginning" to "expert" or "mastery," with any number of levels in between.

This rubric again points to the critical nature of faculty ownership of the assessment process in setting expectations for students' learning and performance. The understanding that students are consistently performing at the "beginning" level when we are urging "mastery" tells us where to start the conversation about our curriculum, co-curriculum, and other points of influence that should be the focus of review and revision.

THE POWER OF CONNECTING TECHNICAL COMMUNICATION ASSESSMENT TO INSTITUTIONAL VALUES

Of course, the connection between institutional values and the technical communication program is just the first step in a valid map of the curriculum and the technical communication faculty's expectations. More discipline-based learning regarding genres, processes, and products would certainly be articulated in the remainder of their assessment work. But it is nonetheless valuable to add this "mile-high" view of the connection to the institutional values in our thinking about our programs and assessing their impact on our students, especially as we work to articulate how their degree from our institution differs from a technical communication degree from another institution.

In addition to providing a grounding framework for assessing the technical communication curriculum, the use of institutional values as a context for such assessment offers other advantages. Foremost may be the evidence that the curriculum is a key provider of the institution's stated values. Few disciplines, in fact, can lend themselves to institutional values and contexts more readily than communication, since it necessarily exists in every context, every curriculum, and, in some fashion I would even argue, in every desirable outcome.

Second, because every institution faces the realities and consequences of scarce resources, every curriculum's faculty and directors seek evidence of the contributions or even the centrality of their programs to the institution's core values. That evidence is a powerful mechanism for bringing attention and resources to the technical communication program from a number of sources, including standard internal [re]allocations. Quite simply, while new ideas and programs appeal to the entrepreneurs and visionaries among us, senior administrators do not get to enjoy that privilege if they have not first protected and strengthened existing core programs.

Just as important, however, is the attention and resources that can come to successful programs from donors and granting foundations. To put it a bit crassly, donors like to back winners: they prefer to give their money (or other resources) to successful programs or, at the very least, to programs that can not only articulate a vision but demonstrate a strategic plan for increasing the likelihood of success. And granting agencies and foundations assume greater strength in academic programs that are clearly tied to the core values of the institution. The kind of visibility that evidence of success can provide is invaluable in moving the technical communication curriculum into the spotlight and, thus, into the line of funding.

Finally, but hardly least important, is the clarity that such a context can provide for faculty and students. Competing urgencies bombard curricular planners at every turn: Should we incorporate more technology into the curriculum? Should we focus on better or higher-paying internships? Should we expand our program? Narrow it? Make it more or less exclusive? Should we try to attract a different kind of student? Or a different kind of faculty?

Knowing the values of the institution and the role of the technical communication program within those values may do one of two things: (1) help articulate or stabilize the priorities of the program or, alternately, (2) help move the institution's values in a new direction. As faculty members are increasingly aware of the need to articulate or question institutional priorities that distinguish their institution from its competitors, they should be more effectively engaged in conversations that shape the institution's future, such as a change in mission (e.g., social justice, civic engagement, experiential learning), curricular or pedagogical focus (e.g., technology-based, service-learning, inquiry-guided learning), or student population (e.g., adult learners, distance learners, learning disabled, multi-ethnic). While any change in mission or values is highly complicated and, in some circles, practically impossible, the opportunity to understand what our institutional values communicate about our understandings of the world our graduates will enter—and the specific role that our own institution can play in preparing students for that world—is one of the most important conversations we can have in higher education. For technical communication faculty to be prepared for that conversation, whether leading it or not, is critical to the advancement of our discipline as a primary site for learning in general and for learning the particulars of knowledge needed for success (however we may define it) in the 21st century.

Criterion	Beginning	Moderate	Mastery
Developing sophistication of the writing in terms of fluency with style, rhetorical strategies, design, and other features.	Student demonstrates little to no understanding of alternative ways to express ideas or ways that those alternatives affect the message; student demonstrates little to no understanding of the differences between various rhetorical strategies and has limited capacity to recognize or alter prominent rhetorical features of a document; student can recognize the visual features of a document but lacks the background to understand defensible (i.e., well-researched) selections of one feature versus another; student is likely to seriously over- or under-use design elements in documents.	Student shows ability to recognize stylistic, rhetorical, design, and other features in the works of others and is capable of articulating the impact of those features on the document itself and on the audience's likely understanding of the work. Student is increasingly able to transfer that knowledge and ability to his/her own writing at either the invention or editing stages of the work. The student can create alternative versions of a document to demonstrate his/her fluency with choices. Student can reasonably articulate the likely impact of the alternative versions and make an appropriate selection for the final version of the document. Student has at least some understanding of the evidence that research and informed theory provide for those decisions.	Student can readily analyze the key features of a writing situation and weigh alternative stylistic, rhetorical, design, and other communicative elements. Student can move easily between alternative versions of a document and make appropriate edits and decisions that culminate in a superb final document. The student can articulate not only the rationale for his/her choices, but also the theoretical and/or research-based underpinnings that justify that selection.
Fluency with genres	Student makes little to no distinction between various genres of writing, either in understanding the purpose or use of the genre or its characteristic features. He/she has little to no recognition of the decisions that culminate in the selection of a	Student recognizes the variety of genres available for the fluent writer. He/she is able to recognize and describe the key features of most genres. He/she is able to work within a genre's expectations to create a fluent document that reflects	Student is highly capable of articulating the key features of various genres and using those features as a valuable scaffold for building his/her own document. Student can make substantive decisions about the appropriateness of modifying that

	particular genre and, then, its concomitant features.	at least the critical features that define a genre. Student has at least some understanding of the research and informed theory that supports the identification and creation of a genre.	scaffolding for the particular content, context, audience, or purpose of the document. Student has a well-researched understanding of the history, key features, and theoretical underpinnings of genres.
Connectivity between content, context, audience, purpose, and other features	Student demonstrates modest understandings of the critical connectivity between multiple facets of the writing situation. Student focuses on only one or two elements in a document's creation. Student may begin to show signs of making individual considerations of each element but cannot weave together their fluidity in a sense of connectivity.	Student recognizes the full impact of the connectivity between content, context, audience, purpose, and other features. Student manages at least the major features in documents, according to the individual writing situation, with some appropriate level of consistency. Student can make and defend choices, based on considerations for creating a well-connected document, with some level of consistency.	Student easily recognizes the critical interplay between content, context, audience, purpose, and other features. Student can readily identify the critical complexities of that interplay, especially where one or more of the elements conflict or complicate the assignment. Student has a highly developed repertoire of strategies for overcoming the complexities, resulting in the creation of a well-connected document.
Conventions of standard written English (or equivalent language)	Student has little to no consistent ability to apply the conventions of standard written English. His/her work is riddled with grammar and punctuation errors that he/she is unable to explain or correct.	Student writes relatively flawless English, with only modest and then, most likely, minor errors. Student is typically able to understand and articulate the conventions as he/she corrects the document.	Student can readily apply the conventions of standard written English to any document or project. Student can easily explain the convention and demonstrate its value in promoting clarity of the work and respect for the reader.

Figure 3. Rubric of portfolio evaluation criteria for level of competency.

REFERENCES

Anderson, J. (2004). An institutional commitment to assessment and accountability. In P. Hernon & R. E. Dugan (Eds.), *Outcomes assessment in higher education: Views and perspectives* (pp. 17–28). Westport, CT: Libraries Unlimited.

Astin, A. W. (1993). *Assessment for excellence: The philosophy and practice of assessment and evaluation in higher education*. Phoenix, AZ: Oryx Press.

Huba, M., & Freed J. E. (2000). *Learner-centered assessment on college campuses: Shifting the focus from teaching to learning*. Boston, MA: Allyn & Bacon.

Nichols, J. O., & Nichols, K. W. (2005). *A road map for improvement of student learning and support services through assessment*. New York: Agathon Press.

CHAPTER 4

The Benefits and Challenges of Adopting a New Standpoint While Assessing Technical Communication Programs: A Response to Jo Allen

Paul V. Anderson, Miami University (Ohio)

In "Mapping Institutional Values and the Technical Communication Curriculum" (see Chapter 3 in this volume), Jo Allen has created the best kind of assessment discussion, one that tells us how to use assessment to strengthen our academic programs through thoughtful, well-designed, evidence-based action. Furthermore, by looking at the assessment of technical communication programs in a novel way, she identifies a new assessment-related strategy for faculty and program directors.

A FOCUS ON OBJECTIVES

In higher education, activities called *program assessment* are used for two distinguishable purposes. Both involve gathering data or other evidence to determine how well a program is succeeding at achieving its objectives. The first purpose is to judge whether the program measures up to a predetermined level of achievement. A program that measures up may be praised and, perhaps, rewarded. One that doesn't may experience many fates, including mandates for change, reduction in support, or dissolution.

In her chapter, Allen focuses on program assessment conducted for the second purpose: to identify areas in which the program can be improved. In this type of assessment, improvement (rather than judgment) remains the goal even for programs that more than measure up to whatever standards apply. Its final step is often called "closing the loop," because the knowledge gained by assessing the program is used to refine the program. Once the modifications have been completed, the assessment process may begin again, looking for additional areas for improvement. Such ongoing assessment is generally depicted as a cycle, such as the one shown in Allen's Figure 2, The Assessment Cycle.

The originality of Allen's chapter springs from her decision to focus on a different point in the assessment cycle than is customary. The literature on assessment in higher education is replete with advice about the kinds of data or evidence to collect, ways to analyze it, and ways a program could act on the results. Less plentiful are discussions of the ways a program can identify the objectives that are most worthy of pursuing. In technical communication programs, as Allen points out, we generally construct our objectives by consulting several sources, including our faculty's interests and knowledge of our field, the needs of the employers who hire our graduates, and the objectives adopted by programs at other institutions. No one else has suggested that we consult the source to which Allen draws our attention: the values of the institutions in which our programs are located.

DEFINING "INSTITUTIONAL VALUES"

The power of Allen's advice derives from the manner in which she defines *institutional values*. Before I read her chapter, the term evoked for me the type of values affirmed in the official "values statements" adopted by the trustees of many institutions, including mine, Miami University (Ohio). The values proclaimed by Miami's statement (2002) include—but are not limited to— honesty, integrity, moral conduct, and respect for the dignity and property rights of others. In contrast, in a passage I find to be particularly thought provoking, Allen associates an institution's values not with such abstractions such as honesty and integrity but rather as "desired outcomes." As example outcomes, she names characteristics of the school's graduates: "civic leaders, global learners, citizens of character, industry leaders." To me, this passage suggests that we might think of institutional values as the valuable things a college or university gives back to society. "Entrust us with your young adults and other citizens," we might say, "and we will give you back graduates who are civic leaders, citizens of character, and other highly desired individuals."

Of course, our institutions give society many other valuable gifts, such as research findings that propel industry and advance medicine, cultural events, and sports entertainment. However, none is a more valuable outcome of our educational enterprise than the valued persons our graduates will be. By suggesting that we consider institutional values when we define the attributes we want our technical communication graduates to possess, Allen challenges us to think beyond the specialized, professional abilities they will use in their careers. Even in our professional, practice-oriented programs, we should explicitly and intentionally strive to imbue our students with the general attributes that our institutions want all of their graduates to display.

As Allen points out, colleges and universities may express their institutional values—the values that define the qualities they aim to instill in their students—in many places. We may discover these values in a formal values statement, an institutional mission statement, or an unrecorded but palpable agreement among faculty, students, administrators, and staff. Often, institutional values of the kind Allen discusses are expressed in the goals of the general education requirement. Such is the case where I teach. Here, the general education requirement emerged from several years of discussion among faculty. It was adopted by University Senate. Rephrased as the kind of "desired outcomes" that Allen discusses, this requirement affirms Miami's desire to give society graduates who think critically, understand issues and problems in terms of their multifaceted contexts, engage with others on important issues, reflect on what they learn, and act in accordance with their knowledge and ethics. None of these habits of thought are inimical to the professional skills taught in a technical communication program, but they are not likely to be highlighted in it either.

MATERIAL AND INTELLECTUAL BENEFITS OF CONSIDERING INSTITUTIONAL VALUES

To persuade us to incorporate our institution's values into the objectives of our technical communication programs, Allen emphasizes the material benefits of doing so: Our programs may gain more internal money, more faculty lines, and more donations from persons and organizations outside the institution. Her observations take on special force because of her substantial administrative experience, including work as an Interim Vice Provost at North Carolina State University and Provost at Widener University. Coupled with her longstanding devotion to our field, this administrative experience gives Allen a double vision of our technical communication programs. She knows not only how we would like our programs to be viewed but also understands the perspectives from which academic administrators view all programs, including ours.

In addition to enhancing our programs' chances of enjoying material benefits, I believe we can increase the intellectual strength of our offerings if we incorporate institutional values, as Allen defines them, into our program objectives. For example, while reading Allen's chapter, I began to consider ways the values expressed in Miami's general-education requirement might enrich Miami's undergraduate and graduate programs in technical communication. I have felt very adept at explaining how the programs help students develop the attributes whose value is asserted in the university's general-education requirement. Miami's institutional value of critical thinking provides an example. Critical thinking, I would explain, involves seeing an idea or situation from other points of view than your own. Our technical communication programs inculcate this ability in many ways, I would continue, by teaching audience analysis, explaining ways to identify and address counterarguments, and discussing ways to investigate the ways a communication could impact all individuals who might be affected by the action it advocates. However, reading Allen's chapter prompted me to revisit Miami's description of critical thinking. It includes, I was reminded, the act of examining the assumptions that underlie the positions others take and also those we ourselves take. I now see that I could provide a richer education if I addressed this aspect of critical thinking, something I would not have focused on without following Allen's advice.

INCORPORATING INSTITUTIONAL VALUES INTO PROGRAM OBJECTIVES

To incorporate institutional values into the assessment of technical communication programs, we must restate them as educational objectives that can be assessed. Allen shows us how to do this by drawing on current practice within the assessment movement in higher education. Institutional values, she explains, are framed as "desired outcomes" or "student learning outcomes." Allen's Table 1 illustrates the process. The institutional goal of providing students with "written communication skills" (first row) is translated as a series of specific writing tasks, called "general outcomes" (second column), that the institution aspires to teach every student, regardless of major, to do before graduation. Next, general outcomes are translated into more specific capabilities that are expressed as the "student-learning outcomes" for the technical communication program (third column). Thus, in one of Allen's examples, the institutional value of "communication skills (written)" is operationalized as a series of general outcomes that include the ability to "explain common disciplinary issues . . . to expert and lay audiences." In turn, this general outcome is incorporated in the technical communication program's desired outcomes as the ability to "translate any subject to any group of readers, ranging from experts to managers, to special interest audiences, to lay readers." By examining student projects written in response to assignments that ask them to create communications for various audiences, a program's success at achieving this student-learning outcome can be assessed.

Of course, translating the institutional value of "communication skills (written)" into an outcome for a technical communication program seems straightforward to us. Nevertheless, there can be a substantial benefit to considering institutional values in the way Allen urges. This consideration can help us spot important professional and intellectual skills that we should have included in our program's objectives but have overlooked, just as I overlooked the program objective of teaching students to examine underlying assumptions.

CHALLENGES

To fully realize the advantages of incorporating institutional values into our lists of outcomes that we will assess, a technical communication program may need to address several substantial challenges. To begin, institutions have many "desired outcomes." Identifying the ones most likely to afford advantages to our programs can be difficult. I identified the ability to examine underlying assumptions as a desired outcome for Miami's technical communication programs because that ability is part of Miami's written description of "critical thinking," a cornerstone of its general-education program. However, any institution, including Miami, may strongly desire some outcomes it has not stated in writing. Or, the written statements may be buried in less-than-obvious places, such as presidential addresses, university marketing materials, and the like. Moreover, programs are valued and resources are distributed differentially among them by individual administrators. Consequently, the outcomes a program can most advantageously address are the ones that are most desired by particular persons. As one administrator is replaced by another, the outcomes most desired by the "institution" may also change. They can also change during the tenure of one person. With plenty of justification, a president or provost may alter his or her list of highly desired outcomes in response to changes in the social and political climate for higher education, new members on the board of directors, or new ideas for higher education advanced by such organizations as the American Association of Colleges and Universities.

Once we've identified them, some of a university's desired outcomes may be difficult to define in assessable forms. Allen identifies "citizens of character" as one desired outcome. Similarly, Miami University's "Statement of Institutional Values" cites "moral conduct." How would an institution create assessable definitions of such terms? Any operational definition is likely to provoke controversy among faculty and administrators. Even if a university were able to develop widely accepted and assessable definitions of these terms, it would have difficulty determining how much any particular program contributes to student gains in such areas. Students' development in them probably results from the convergence over an extended period of all of their in-class and out-of-class experiences, not just their studies in a specific program.

Finally, we would face the challenge of designing assignments that enable us to assess our programs' effectiveness at helping students achieve desired outcomes that are based on institutional values. We can easily assess the extent to which our programs succeed in achieving disciplinary objectives. For example, when students prepare instructions, proposals, reports, and Web sites, they demonstrate their abilities to design effective pages, adapt content and style to a particular reader, present data in easy-to-use graphics, and so on. When they turn in these assignments to satisfy course requirements, they are also giving us the artifacts we need in order to assess our programs' effectiveness at achieving discipline-based learning objectives. No extra activity is inserted just for the assessment's sake. Walvoord (2004) and others would say that this kind of assessment is "embedded" in our programs. It would be more difficult for us to embed assessment of many learning objectives based on such institutional values as the development of "moral conduct." Many other institutional values (or "desired outcomes") present us with a similar challenge.

CONCLUSION

The literature on assessment provides an abundance of advice for planning and implementing most phases of the assessment cycle, especially the phases of analyzing student artifacts and refining curricula based on what is learned. Far less advice addresses the fundamental task of defining a program's educational objectives. Allen has increased our ability to improve our technical communication programs by suggesting that we use institutional values as one source for our program objectives. By doing so, we may attract additional material support for our efforts even as we discover ways to create a more robust curriculum and a richer set of aspirations for our students. Whatever challenges we encounter along the way, this effort is well worth making.

REFERENCES

Miami University. (2002). *Miami University values statement*. Retrieved April 5, 2008, from www.miami.muohio.edu/documents_and_policies/values.cfm

Walvoord, B. E. (2004). *Assessment clear and simple: A practical guide for institutions, departments, and general education*. San Francisco, CA: Jossey-Bass.

A Role for Portfolios in Assessment

CHAPTER 5

Politics, Programmatic Self-Assessment, and the Challenge of Cultural Change*

Kelli Cargile Cook, Texas Tech University
Mark Zachry, University of Washington

Programmatic self-assessment in professional and technical communication presents a dilemma for the faculty involved because the unit of analysis is much different from what we as individual instructors are conditioned to think about, such as assessing the performance of an individual student, the success of an assignment, or the success of an entire course in a given semester. When assessment focuses on our teaching and the learning activities we facilitate—or even when it focuses on students in our individual classes—we have some degree of confidence in the long social history that authorizes such evaluations. That is, instructors are *supposed* to be accountable for their instruction and, in a different and more complicated sense, instructors are accountable for facilitating and measuring student learning. When asked to begin evaluating individuals and activities that we as individuals are not accountable for, however, the proposition is more troubling. Specifically, recently hired and untenured faculty are likely to feel particularly ill at ease when they are asked to participate in—or even lead—a formalized process of programmatic assessment wherein they will appraise their immediate colleagues' instructional practices and the work their students produce. More experienced faculty may, likewise, find initiating the practices of programmatic self-assessment awkward as these practices begin focusing new attention on comfortable assumptions that have gone unquestioned for several academic years. As two faculty members who began their academic

*Both authors contributed equally to this chapter. Zachry worked at Utah State University from 1998 until 2006, Cargile Cook from 2000 until 2009.

65

careers in a program where self-assessment practices were evolving, we can attest to these varied responses from individuals involved in different ways in the process. We, for example, experienced the politics of change in different ways at different times because over the course of several years we were both newly hired faculty members in the program and then later tenured professors, chairing the program's curriculum committee.

In large part, the conversations initiated by programmatic self-assessment seem difficult because our field lacks scholarship that considers how programmatic/ faculty participant identities are shaped in curriculum-committee discussions. As a field, we have little widely shared knowledge about these discussions, which occur behind closed doors and often entail disagreement and dissensus, and are rarely revealed publicly: No one wants to display "dirty laundry" for others to witness and criticize. However, Clemson University's writing faculty's recent chapter, "Notes toward 'Reflective Instrumentalism': A Collaborative Look at Curricular Revision in Clemson University's MAPC Program" (Yancey et al., 2004) begins to reveal the formerly closed negotiations and renegotiations that occur at critical moments in program development and change. Clemson's story of revising its comprehensive reading list illustrates some of the tensions and compromises that may arise when faculty members engage in reflective discourse about their programmatic components. Its honest description of turf wars that occur during reflective engagement in curriculum committees reveals why few have chosen to write about these negotiated spaces and occasions. Yet, like our colleagues at Clemson, we believe that engaging in reflection about our program and its goals, while sometimes difficult, has resulted in a clearer identity of who we are as individual faculty members, who we are as a collective faculty, and what defines us as a distinctive program.

For these reasons, in this chapter, we describe the behind-the-scenes reflective discourse that has shaped and continues to shape the identity of the professional and technical writing program at Utah State. We address the inherent difficulties of programmatic assessment or, more accurately, programmatic *self*-assessment. We begin by discussing a vehicle for facilitating programmatic self-assessment: student portfolios produced in a capstone class during the students' senior year. These portfolios have not only become the agents of our reflective discourse but also artifacts of these discussions. We then expand the scope of this chapter to examine how our annual assessment of these portfolios is always ultimately a self-assessment practice that leads to sometimes productive and sometimes unresolved discussions about the curriculum and related instructional practices. As our study reveals, in the more or less egalitarian culture of higher education, wherein individuals place a premium on their latitude to think and act as individuals, self-assessment processes and results must be calibrated to fit the culture. Over time, these processes and results may also begin to shape the culture in which they are employed.

So, rather than offering rubrics or specific recommendations for how others should use portfolios as part of their self-assessment activities, this chapter has a different focus. It describes how the integration of an annual, collective faculty review of portfolios produced by graduating seniors can beneficially foreground political-cultural issues in a program. Readers who are interested in learning more about specific practices that may lead to programmatic self-reflection will find sources cited in this chapter—as well as other chapters in this collection—a valuable starting point. That is, whereas these other sources offer suggestions about how to engage in self-assessment activities, this chapter illustrates the sorts of political-cultural discourse that can accompany such activities and suggests some of the benefits of engaging in this type of discourse.

STUDENT PORTFOLIOS AS VEHICLE FOR SELF-ASSESSMENT

The professional and technical writing program at Utah State University has been engaged in programmatic self-assessment for the last 7 years, using student portfolios as the primary focus of its evaluations. In the spring of 1997, the program's curriculum committee decided to use seniors' final class projects— professional portfolios that they developed to enter the job market—as assessment data. This decision acknowledges that traditional measures of programmatic strengths (class sizes, technology resources, etc.) overlook the most fundamental issue in assessing the quality of student instruction: demonstrable skill and knowledge in communication that our majors in professional and technical writing should possess. Achievement in communication skills and knowledge beyond demonstration of codifiable practices (e.g., grammar and mechanics) are, of course, difficult to measure, which helps explain why they have not been traditionally included in institutional assessment plans. Nevertheless, at the beginning stages of our program assessment, we assumed that student-produced portfolios were not only objects by which to evaluate students' writing and assign a grade but also evidence of the effectiveness of that student's educational experiences within our program. Consequently, the portfolio could be examined as an artifact that demonstrates where, how, and even to what degree the programmatic curriculum has served its students. Assessing all portfolios produced by seniors over an academic year provided the curriculum committee with a basis for making decisions about the strengths and weaknesses of the students and our program. This added perspective was the key reason why we added portfolio assessment to our self-assessment practices and why other programs might consider doing so also.

Portfolios have been used to measure student performance and curricular effectiveness in disciplines that employ pedagogies comparable to those used in technical and professional communication curricula. For example, scholars in writing program administration (Black et al., 1994; Condon & Hamp-Lyons,

2000; McDonald, 1996; Mullin, 1998; Wolcott & Legg, 1998; Yancey & Weiser, 1997), communication studies (Aitken, 1994), computers and writing (Yancey, 1996), and teacher education (Barton & Collins, 1997; Campbell et al., 2000; Hus & Bergeron, 1997) have all reported their use of portfolios for these purposes. (For a detailed bibliography on portfolios, see Rebecca Moore Howard's (2004) Web-based list of sources at http://wrt-howard.syr.edu/Bibs/Pfs.bib.htm).

Our focus on self-assessment differs substantively from more common uses of portfolios in professional and technical writing programs, such as assessing individual student writing performance (Allen 1993; Bishop, 1989; Elliot, Kilduff, & Lynch, 1994; Senf, 1983) and promoting the students' first professional job searches (Powell & Jankovich, 1998). Instead, our focus is more in keeping with Dillon (1997), who discusses using student portfolios for technical and professional writing programmatic evaluation, indicating that portfolios can be viewed as indicators of programmatic achievement and that they "provide a touchstone for important and ongoing dialogue about what works and what doesn't in technical and business writing" (pp. 42–43). However, in Dillon's study, students voluntarily enter a portfolio contest to be judged by advisory-board members, which makes this use of portfolios quite different from our own.

Including an assessment of student portfolios in our programmatic self-assessment mix did not seem unusual or difficult to us, although the faculty recognized that an annual assessment of these capstone portfolio projects would be time-consuming. From the beginning, the faculty accepted this responsibility along with other assessment practices without much discussion or dissension. What became more difficult was the practice of these evaluations and our application of what we learned from them. We found rather quickly that we could define methods to evaluate the portfolios, producing data that would suffice for our mandatory annual report; but we learned within the first year of implementation that while gathering and reporting data to higher administration was easy, making it meaningful to us as a curriculum committee was much more difficult. Thus, we began refining our processes for developing and evaluating portfolios for program assessment—reworking our portfolio assignment, our measurement tool, and, eventually, our entire curriculum along the way.

ASSESSING STUDENT PORTFOLIOS AND THE POLITICS OF SELF-ASSESSMENT

Over the years we have been assessing student portfolios in order to assess our program, we have become acutely aware of factors that frame what assessment is and what it can do. Before discussing these factors below, we briefly review the 7-year evolution of our using the portfolio as an instrument of program self-assessment.

The Evolution of the Portfolio Assignment

As we write this chapter, the portfolio assessment process at Utah State has been in place for 7 years. Over the first 3 years, the portfolios we assessed were exact copies of capstone students' professional portfolios, which contained six to ten examples of professional quality writing as well as introductions to these examples. To differentiate the two portfolios, we call the job-search portfolio the students' "professional" portfolio, while the portfolios that students create for faculty to assess the program have become known as "academic" portfolios.[1] In the first 3 years, the document samples in both portfolios were frequently the same. Each document's introduction contextualized its audience, purpose, and the technologies used to produce it.

In years 4 through 7 of the process, two significant changes took place, driven in part by the arrival of new faculty with research backgrounds in rhetoric and technical communication pedagogy and by new assessment requirements from university administration. The first change was made in the document introductions in the two portfolios. The professional portfolios continued to include single-page introductions for individual pieces, but the academic (program assessment portfolio) introductions were enhanced with an additional section in which students discussed the program objectives that were demonstrated in each document. Our intention in making this change was to help students develop a stronger understanding—through their writing in this portfolio assignment— of the overall coherence of the program of study they were completing. This change was developed and implemented in the 5th and 6th years. It was accompanied by changes in the portfolio assessment tool, which we discuss later in this chapter.

In the 7th year, we implemented another significant change in the portfolio assignment. For a variety of reasons, including saving student resources and making storage of the academic portfolios easier, the assessment committee recommended that the academic portfolio be delivered to the faculty as electronic documents, created in HTML and presented to the faculty on CDs. This change required students to demonstrate their facility with digital design and Web development. At the same time, the introductions to the academic portfolio were returned to the original context-setting focus used in the professional portfolio, but an extended, reflective essay demonstrating student knowledge and application of program objectives was added as a new requirement for the academic portfolio. These changes streamlined portfolio development for senior students in the technical communication capstone course. Once again,

---

[1] For a detailed discussion of the differences in the academic and professional portfolio assignments and the introductions employed in them, see Cargile Cook's (2002) "Layered Literacies: A Theoretical Frame for Technical Communication Pedagogy."

the academic and professional portfolios were virtually the same documents except that the academic portfolio was presented on a CD and included one additional document: the reflective essay.

In the upcoming year, we will assess this variation of the academic portfolio for the first time to determine its usefulness and effectiveness as a means of programmatic self-assessment. Throughout this 7-year process, we have become increasingly aware of factors that complicate it. Of these factors, three have promoted discussions and affected our programmatic culture most dramatically: the process of developing a common set of ideas and vocabulary with which to define our programmatic goals, the development of self-regulating behaviors among our faculty, and the continual flux of program development and the changes this fluctuation requires.

Developing Common Ideas

We have discovered that high-level program objectives and generic course objectives make agreement easier but also weaken self-assessment practices—if those practices are understood to mean that there is consensus about the specific things students should learn. In other words, the more generically we state our programmatic and course objectives, the more likely we are to have consensus about them, because generic objectives allow more autonomy and room for personal teaching preference. Defining the objectives more specifically, in terms that can be measured or observed for assessment purposes, however, reduces flexibility and limits preference, resulting at times in dissensus among the faculty. We have discovered, however, that such dissensus can often be translated into programmatic plusses.

Defining objectives that can be observed or measured in student portfolios has led us to appreciate and expect a certain level of dissensus among faculty members, which both enriches and extends our understanding (as a faculty) of the many representations programmatic goals can take. Such dissensus allows faculty members to employ diverse teaching and learning approaches with students. These approaches are evident when our students discuss what they have learned about document-design practices and job-search materials, such as résumés—topics about which our faculty have widely differing opinions. Exposing students to differences of opinions, especially opinions grounded in research and practice, provides them with a diversity of experiences and knowledges from which to draw when they leave their academic lives for the workplace.

Our agreement to disagree about certain definitions of programmatic objectives as well as methods for their achievement would have been moot, however, had we not first engaged in a process of articulating common ideas through extended discussions of programmatic goals.

Struggling to Define Common Terms for Programmatic Goals and Objectives

Our conversations about programmatic goals and objectives are probably best represented in our portfolio evaluation instruments. In the first iteration of this instrument, portfolio assessment focused on students' professional skills and examined whether students' portfolios contained a variety of documents that were professionally revised and presented in the portfolio itself. As a whole, portfolios were assessed on five criteria: overall presentation (legibility, introduction contents), graphic content (graphic choice and use of color), document size (length, text features), publishing format (electronic or print, accessibility), and purpose. These criteria reflected faculty consensus regarding program goals as well as their common understanding of the repertoire of skills employers seek in graduates entering the workplace.

The addition of new faculty members and their fields of expertise brought changes to the conceptualization of programmatic goals. These changes were exemplified in the original portfolio assessment instrument with its emphasis on both print and digital documents (a change effected by the addition of a faculty member who brought computer-based instruction to the program). As other new members joined the program, their influence was equally as evident, adding more specific focuses on rhetorical, social, technological, cultural, and ethical instruction. In the 3rd and 4th years of the process, faculty discussions became infused with new terminology, and the new faculty members were engaged in revising the portfolio assessment instrument to reflect this broader set of competencies (or as we called them, "literacies").

Although these terms were accepted, in general, as good indicators of student performance in the program, using the new assessment instrument was a process fraught with difficulties, mostly arising from differences in opinions about what terms meant and how they might be demonstrated in student work. Two concepts exemplify the sometimes heated discussions that erupted as we struggled to arrive at a common articulation of program goals. Among the terms most divergently defined was the one that faculty members used to describe the individual or individuals who need or use a professional and technical writing document. The terms faculty members used to identify this individual or these individuals reflected their own teaching specialties and expertise. For example, the faculty member who taught proposals from a practitioner's perspective wanted the portfolio assessment instrument to call these individuals "customers" or "clients"; faculty members from a more traditional academic background preferred "audience," "reader," or "listener"; other faculty members with expertise in usability methods preferred the term "user." No one was particularly willing to give up his or her preferred term. In the end, the assessment instrument reflects a compromise, identifying "client/user/customer" as signifiers of these individuals instead of employing a single term to represent them.

A second concept that provoked extended discussion was "ethics." Some faculty members, who came to technical and professional writing primarily through professional experience rather than academic specialization, defined ethics from a viewpoint focused most directly on "ethics in writing," viewing ethics as primarily expressed in programmatic instruction to students about research and documentation strategies, avoidance of plagiarism, and appropriate citation styles. Other faculty members took a much broader view of ethical instruction, including in it instruction about document stakeholders, organizational codes of ethics, the ethical use of software, and environmentally friendly choices in document design and publication. In many ways, these differences in definition and interpretation reflected faculty members' backgrounds, experiences, and education as well as their teaching preferences. For some, the suggestion that ethics should be so broadly defined was unacceptable and an unnecessary appropriation of teaching time; for others, it was unacceptable *not* to include ethical discussions in their courses. In the end, this concept remained as a category on the portfolio assessment instrument, but how faculty members engaging in assessing student portfolios determined whether the concept was evident in student work remained fluid and varied, as did how or whether faculty members chose to incorporate ethical discussions and instruction into their course curricula.

Demonstrating Program Objectives through Portfolio Development

Agreeing (to some extent) to a more broadly defined set of programmatic goals and articulating how we expected to see these goals exhibited in the portfolios was just the first step in our assessment process. To fully implement this step, we needed to review the previous year's collection of portfolios to see how well the instrument would ascertain students' achievement of these goals. During our review, we realized rather quickly that some program objectives were easy to assess (for example, basic writing skills such as punctuation and grammar skills), while others were virtually invisible in the artifacts contained in a given portfolio (e.g., students' research processes and ethical practices). This realization led us to add the reflective component to the portfolio assignment. We needed students to tell us what they had done to consider all the stakeholders in specific documents; we needed them to describe the primary and secondary research processes that preceded the actual writing of the document; we needed them to critique their documents, telling us what they had done well and what they would change were they to do it all again. Without student insight and foregrounding of their processes, what we considered to be important knowledge domains remained invisible. We could only guess at their achievement.

Engaging Students in the Programmatic Assessment Process

Rather than change the portfolio assessment instrument, eliminating these less evident or less easily demonstrated program objectives, in the 5th year we chose to engage students in the process of articulating their literacies, hoping that they could tell us or show us how they achieved program goals. Capstone-course instructors taught students what the faculty meant by "literacies" and asked them to tell us, in their introductions, when and how their documents demonstrated these literacies. In reviewing this class's portfolios in the 6th year, we were pleased to discover that students could articulate and describe their achievement in many of these objectives for us, and evaluation scores in some objectives rose dramatically from the 2001 to the 2002 scores. For example, on a 4-point ascending scale, "insightful reflection on past writing experiences" rose from 1.45 to 3.08, and "understanding of client's/user's/customer's role in shaping effective discourse" rose from 2.56 to 3.46. Interestingly, however, with some objectives, the reflective essays showed us that students were not achieving them as well as they should be: "understanding ethical issues" dropped from an average of 3.31 in 2001 to 2.00 in 2002, and "success in varied roles on collaborative teams" dropped from 3.02 in 2001 to 2.14 in 2002.

To further improve our students' understanding of the program goals, we decided, following our 6th year review, to introduce students to these goals in the two introductory courses to the major, revisiting them again extensively in the final capstone course. We will evaluate the portfolios from this group of students in our upcoming 8th year.

Changing the Program Objectives and Instruction

Over the course of the past 7 years, this ongoing process of defining common programmatic objectives or goals (or, in our case, literacies) has helped shape our program's contents, its organization and instruction, and our own teaching methods. As individual instructors, it influences how many of us think about our courses. For example, in the 7th year of the process, we revised our published course objectives to include more specifically the same language that we used in the portfolio assessment. It has also begun to shape the way we talk to our students about what they are learning. As a group, we have decided to use our common vocabulary for program goals extensively in three courses—the two introductory courses and the capstone course. In the courses in between, many professors are adding more reflective essay assignments to improve students' critical thinking about their own work, and each of us has committed to assigning a portfolio-quality document in each course, although what each of us defines as "portfolio-quality" is not always the same. In effect, the adoption of a common vocabulary for articulating and evaluating program goals has changed the way we evaluate our students within courses, and it has changed the way our students talk about what they are learning. Although we continue to struggle with certain objective terminology, their meanings, and their demonstration in student portfolios, we accept this struggle as part of program assessment practice and value it as a means to clarify our own understanding of the field and its various representations.

Self-Regulating Behavior

While developing a shared language for discussing assessment presents a set of ongoing challenges, so, too, does creating a culture in which the participants voluntarily regulate themselves to make the practices of assessment meaningful. At its core, programmatic assessment is predicated on the idea that once deviations from a shared set of standards have been observed, the instructional practices that yielded those deviations will be modified so that the agreed-upon standards can be realized. In practice, however, the challenge of getting faculty members to change their instructional practices can be tremendous. In large part, this challenge is attributable to the nature of academic culture, wherein instructional practices—and particularly individual practices of permanent faculty members—are not easily dictated by a central oversight body (Graham, Birmingham, & Zachry, 1999). Except in serious situations, such as individual or faculty misconduct, no external enforcement mechanisms define or ensure implementation of instructional practices that address deficiencies discovered through programmatic self-assessment.

In academic cultures, where individuals are given great latitude for achieving broad instructional objectives, attempting to use a self-assessment process to achieve anything other than standards that have been defined in the broadest of terms is nearly impossible. In short, there is always ultimately an incompatibility between academic culture and a self-assessment process that seeks to continually refine objectives and practices.

In our experience, addressing this gap in any meaningful way depends on the willingness of participants to engage in self-regulating behaviors that will lead to the group's achieving a common goal. We have seen this sort of self-regulating behavior at many key junctures in our multiyear history of programmatic assessment via student portfolios. Instructors, for example, have often elected to share syllabi, assignment sheets, and even grading criteria to ensure that proven practices for achieving programmatic goals are transferred from section to section of a given course. Likewise, instructors have at key moments in this process been willing to abandon a personal favorite topic or assignment in a class that is otherwise essential for achieving a larger, programmatic goal.

Such self-regulating behaviors, however, are by no means always predictable. For nearly every instance in which an instructor has opted to modify her or his instructional practices, we can point to a counter example wherein an instructor has decided to pursue her or his own agenda. For a variety of reasons (only a few of which are identifiable), instructors opt to exercise one or more of the options for resistance available in academic culture to pursue their own ends rather than those articulated by the curriculum committee as a whole. The difficulty of this situation is exacerbated when those with less social capital (e.g., less seniority, lower academic rank, etc.) are more interested in implementing things learned through self-assessment than are those with greater capital who are ostensibly supposed to be working toward the same ends.

In the end, the apparent incompatibility between the largely autonomous instructional culture of higher education and the controlled and monitored end objectives of programmatic self-assessment cannot be erased. At best, this incompatibility can be foregrounded in committee discussions, and individuals can be encouraged to engage in self-regulating behaviors that complement the group's objectives.

Changing Contexts

An additional complicating factor in the activity of self-assessment is that the contexts in which this activity occurs are always in flux. While in the midst of self-assessment activities, it is difficult to remember that the criteria and goals of the assessment are defined in time and space—and that these are relentlessly changing. Excellence in the practice of professional and technical communication does not look the same today as it did in 1995 or 1985, or 1885. Further, identifying where we are situated as instructors, students, and/or practitioners is not possible because we can no longer account for—with confidence—all the possible answers. To illustrate this challenge, we will point to three sites of change within our program that have complicated programmatic assessment based on student portfolios.

First, like many programs in the last 10 years, we have expanded the number of faculty in professional and technical communication. From 1998 until 2004, we hired five new tenure-line faculty in this area. With each new hire, our collective teaching abilities have expanded, and consequently, so has the curriculum. With the addition of teaching expertise in areas such as user-centered research, editing, organizational rhetorics, and new media, the ideas that our students are being exposed to and accountable for have changed dramatically over the last 7 years. Subsequently, we have periodically revised our program objectives and our expectations for programmatic achievement as demonstrated in student portfolios. These shifting expectations become explicit in our instrument for assessing student portfolios, which we have revised three times. Related to these shifts, we have also witnessed a change in our collective expectations for students to demonstrate excellence in areas where we know that new faculty members are providing focused instruction. For example, with the addition of a course dedicated to usability testing and research, the curriculum committee participating in the annual portfolio assessment process now expects that students should be able to demonstrate specific, methodical approaches to audience analysis as opposed to a sort of cursory understanding of "knowing your audience," which was commonplace in student portfolios produced in earlier years.

A second site of change that constantly affects how we assess programmatic achievement as evidenced in student portfolios is the social transformation of communicative practices. As a curriculum committee, a recurring concern we have discussed is ensuring that our instruction remains current in the sense that the topics and technologies that we engage our students with should not seem dated. A curricular area wherein these changes are most evident is editing instruction, which now reaches beyond the how-to approach to traditional editing practices to include discussions of writer/editor relationships, editing for international audiences, and electronic editing strategies and practices. Such changes promote student reflection on their roles as editors as well as their relationships with others within their organizational settings. These changes, however, do not mean that we simply try to reflect exactly the practices of workplaces. Central to the pedagogy of many of the instructors in this area is the idea that informed, critical reflection is essential for student achievement.

A final site of change around which we have made adjustments in our assessment tool is the identities of students themselves. Simply put, we are not encountering the same kinds of students we were in 1997. Students today come to us with much different lived experiences and sets of skills. The students we were working with in the mid to late 1990s, for example, had obviously not long had the World Wide Web as a central venue of social interaction in their lives. Likewise, many students at that time had little or even no experience creating documents with integrated suites of computer software. Today, by contrast, many students bring HTML-coding skills with them to college. Others have been publishing ideas and images via listservs and blogs, or trafficking in the peculiar textual habits of short message service (SMS) for as long as the faculty members they encounter. With these marked changes in our students' identities and skills, our instruction and related assessment expectations have shifted. We expect students to come to us with a certain level of technological literacy, a literacy that now exceeds the highest levels realized by excellent students just a few years earlier. An interesting corollary shift that has occurred with this change in student identities is a return to a stronger emphasis on writing skills. In the mid 1990s, at the time when communication technologies were changing most dramatically, our curriculum focused little attention on advanced writing skills and competencies. Instead, teaching with and about technology eclipsed nearly all other concerns. Now that students arrive better prepared to work with technologies, our instructional and assessment focus is largely returning to excellence in advanced writing and communication design. For example, we offer more instruction in these skills now—reflective writing skills, advanced editing skills, style, and revision.

CONCLUSION

At the most basic level, program self-assessment is a relatively easy task. Identifying program objectives, creating instruments and methods for assessment, analyzing data, and reporting findings with actionable recommendations are all tasks that, in the abstract, do not pose a particular challenge. No one, however, works in the abstract. As faculty, we are embedded in networks of interests and activities that endlessly complicate our work and that prevent us from closing the assessment loop. Assessment thus becomes a perpetual or evolutionary programmatic task that loops us annually through the activities of measuring, analyzing, and recommending program improvements. In our experience, including an annual collective faculty assessment of all the capstone portfolios produced by our graduating seniors has provided a way of thinking about curricular issues in a way that we almost certainly would not have otherwise. In short, this self-reflexive thinking foregrounds political-cultural concerns that would likely go unexamined without such an annual process. And while such collective examinations can be difficult to negotiate, and even while some conflicting ideas remain at the end of the process each year, we believe that, on the whole, our self-assessment practices are more meaningful and productive than they would be without the portfolio component of our self-assessment activities.

In this chapter, we have suggested the sources of key challenges we have faced in the work of assessment at Utah State. Rather than focusing exclusively on specific (and what would ultimately be idiosyncratic) episodes in our 7-year history of self-assessment, we have outlined these challenges in broad terms that are most likely to be applicable in other institutional contexts. It is not difficult to imagine other curriculum committees being challenged in their self--assessment efforts in similar ways. First, language—with its inherent ambiguities—is both a necessary tool for conducting self-assessment and an unavoidable fracture in the plan. Second, self-assessment activities always ultimately depend on the willingness of participants to self-regulate their instructional practices, which cannot automatically be assumed in the culture of higher education. Finally, the practice of self-assessment in any organization is always contingent upon factors (institutional politics, new faculty, new course offerings) that are constantly in play, as is the process of assessment itself.

Given these inherent challenges, the best approach to self-assessment may be to openly and actively engage in discussions along these lines; that is, to draw an explicit distinction between the mandatory practices of self-assessment (e.g., the yearly report) and the factors that are operating beneath the surface (e.g., the ambiguity of key terms in the assessment tool or the sometimes contradictory learning experiences that students are encountering while in the program). While completing the mandatory self-assessment activities is obviously a valuable choice for the curriculum committee as a whole, we believe that so, too, may be discussing in a self-reflective manner the challenges that can never be completely overcome in a self-assessment process. To engage in this self-reflective discussion, we contend, is to become more fully invested in the true spirit of assessment—to create a new curricular committee culture. It moves beyond the mandated, managerial tasks of self-assessment into a shared culture of change, collaboration, and cooperation.

REFERENCES

Aitken, J. E. (1994, August 6). *Assessment in specific programs: Employment, program, and course student portfolios in communication studies*. Paper presented at the Speech Communication Summer Conference, Alexandria, VA.

Allen, J. (1993.). The role of assessment in technical communication: A review of literature. *Technical Communication Quarterly, 2,* 365–388.

Barton, J., & Collins, A. (Eds.). (1997). *Portfolio assessment: A handbook for educators.* Menlo Park, CA: Addison-Wesley Publishing Co.

Bishop, W. (1989, Winter). Revising the technical writing class: Peer critiques, self-evaluation, and portfolio grading. *Technical Writing Teacher, 16*(1), 13–25.

Black, L., Daiker, D. A., & Sommers, J. (1994). *New directions in portfolio assessment: Reflective practice, critical theory and large-scale scoring*. Portsmouth, NH: Boynton/Cook.

Campbell, D. M., Melenyzer, B. J., Nettles, D. H., & Wyman, R. M., Jr. (2000). *Portfolio and performance assessment in teacher education*. Boston, MA: Allyn and Bacon.

Cargile Cook, K. (2002, Winter). Layered literacies: A theoretical frame for technical communication pedagogy. *Technical Communication Quarterly, 11*(1), 5–29.

Dillon, W. T. (1997, March). Corporate advisory boards, portfolio assessment, and business and technical writing program development. *Business Communication Quarterly, 60*(1), 41–58.

Elliot, N., Kilduff, M., & Lynch, R. (1994). The assessment of technical writing: A case study. *Journal of Technical Writing and Communication, 24*(1), 19–36.

Graham, M. B., Birmingham, E., & Zachry, M. (1999). A new way of doing business: Articulating the economics of composition. *Journal of Advanced Composition, 19,* 679–697.

Hamp-Lyons, L., & Condon, W. (2000). *Assessing the portfolio: Principles for practice, theory & research.* Cresskill, NJ: Hampton Press.

Howard, R, M. Portfolios: Some resources. 19 October 2004 http://wrt-howard.syr.edu/bibs/pfs.bib.htm

Hus, L. A., & Bergeron, B. (1997). Portfolios and program assessment: Addressing the challenges of admission to a pre-service program. (ERIC Document Reproduction Service No. ED414257)

McDonald, R. L. (1996, March 27–30). *The writing portfolio and English program assessment: Of bumps, bruises, and lessons learned.* Paper presented at the annual meeting of the 47th Conference on College Composition and Communication, Milwaukee, WI.

Mullin, J. A. (1998, Summer). Portfolios: Purposeful collections of student work. *New Directions for Teaching and Learning, 74,* 79–87.

Polumba, C. A., & Banta, T. W. (1999). *Assessment essentials: Planning, implementing, and improving assessment in higher education.* San Francisco, CA: Jossey Bass.

Powell, K. S., & Jankovich, J. L. (1998, December). Student portfolios: A tool to enhance the traditional job search. *Business Communication Quarterly, 61*(4), 72–82.

Senf, C. (1983, Fall). The portfolio or ultimate writing assignment. *Technical Writing Teacher, 11*(1), 23-25.

Weiser, I., & Yancey, K. B. (Eds.). (1997). *Situating portfolios: Four perspectives.* Logan, UT: Utah State University Press.

Wolcott, W., & Legg, S. M. (1998). *An overview of writing assessment.* Urbana, IL: National Council of Teachers of English.

Yancey, K. B. (1996). Portfolio, electronic, and the links between. *Computers and Composition, 13,* 129–133.

Yancey, K., Williams, S., Heifferon, B., Hilligoss, S., Howard, T., Jacobi, M. et al. (2004). Notes toward 'reflective instrumentalism': A collaborative look at curricular revision in Clemson University's MAPC program. In T. Bridgeford, K. S. Kitalong, & D. Selfe (Eds.), *Innovative approaches to teaching technical communication* (pp. 93–108). Logan, UT: Utah State University Press.

CHAPTER 6

The Road to *Self*-Assessment: Less-Traveled But Essential

James M. Dubinsky
Virginia Tech

In "Politics, Programmatic Self-Assessment, and the Challenge of Cultural Change," Kelli Cargile Cook and Mark Zachry (see Chapter 5, this volume), describe a 7-year journey involving the building and assessing of their program in professional communication at Utah State. Their journey was by no means direct or without construction zones and hazards. To document these zones and hazards, they present some of the "inherent difficulties of . . . programmatic *self*-assessment" [italics in original] (p. 66), highlighting the "political-cultural issues in program[s]" that create or contribute to those difficulties, and suggest some of the benefits of engaging in such *self*-assessment (p. 67). In so doing, they offer a model for assessment and some important questions to consider for those of us who teach and administer programs in the field of professional communication. In this response, I reflect on the problems and benefits that Cargile Cook and Zachry have outlined and offer a few thoughts about ways to extend their work.

PROBLEMS

The question of what makes a writing program successful is a difficult one. As Cargile Cook and Zachry state, scholarship on the question is limited, particularly in our field of professional communication. However, some of the work from the fields of rhetoric and composition and writing across the curriculum (WAC) is useful, particularly the work of Ed White. In *Developing Successful Writing Programs* (1998), White, relying on data from several major studies,

81

outlines a series of features that well-developed writing programs contain, two of which are "agreed-upon and applied criteria for student writing performance at the various levels of coursework" (p. 7) and "an assessment design" (p. 8). Yet, even as he outlines strategies for assessment design, he explains that agreeing upon criteria to evaluate performance is not easy to do because "there is no replicated design in existence for demonstrating that any writing instructional program in fact improves student writing, if we define writing in a sophisticated way" (White, 1998, p. 198).

White recognized two critical issues: the need for agreement among faculty about criteria and key terms (definition), a need which involves creating a community of scholars and teachers dedicated toward common goals. Cargile Cook and Zachry also focus on these issues, suggesting that "political-cultural" issues involved in the process of trying to achieve agreement within a program are serious roadblocks to achieving the kind of replication White sought. Drawing on stories from their program and those of faculty at Clemson (Yancey et al., 2004), they suggest that the way to the end may only be through the turmoil. By describing the turmoil, the process and the tensions involved in *self*-assessment—between theory and practice; between requirements from university administrators to standardize, document, and quantify practice and resistance of faculty members to give up academic freedom; and among varying claims for turf from faculty within a program—Cargile Cook and Zachry *re-create* in their readers, at least in this reader, a sense of the difficulties involved in making assumptions faculty hold about objectives, standards, and assessment explicit. Leaving these assumptions tacit or unexplained typically means that little or no change will occur. Explaining professional knowledge through the kind of dialogue they model, however, makes tacit knowledge explicit and leads to change (Argyris & Schön, 1975; Polanyi, 1967).

As a program administrator, I welcomed their trailblazing. I had just finished an entire year (actually several) with a diverse group of colleagues devising a curriculum for a PhD in Rhetoric and Writing, and I came to this essay with a heightened sensitivity to the cultural and political issues involved in curriculum design and assessment. Thus, I was fascinated to learn about the difficulties and negotiations that occurred at Utah State, and I appreciated the authors' frank discussion about the importance of language, willingness of participants to self-regulate, and other factors that affect assessment (e.g., institutional politics, new faculty, etc.) (p. 70). Their story highlights the value of outlining and defining course and programmatic objectives as essential tasks, and the need for a common language and a common foundational understanding. Their example about the discussion and "dissensus" over the terms "client," "customer," and "user" highlights that need. These terms clearly demonstrate different pedagogical and political positions, different sets of assumptions about how different faculty *see* students and the projects they ask students to complete. Making such a discussion foundational, incorporating it into the planning process, makes much sense, as does engaging students and bringing them into the discussions.

Equally important is their acknowledgment of the "challenge of getting faculty members to change their institutional practices" (p. 74). Cargile Cook and Zachry are accurate when they illustrate the difficulty and unpredictability of getting faculty to sacrifice a bit of freedom; and their model, which illustrates how building community through dialogue is an effective means to achieve some standardization, seems replicable. However, they conclude that "in the more or less egalitarian culture of higher education wherein individuals place a premium on their latitude to think and act as individuals, self-assessment processes and results *must be calibrated to fit the culture*" [italics mine] (p. 66). Perhaps they are too generous here, putting faculty members' freedom ahead of programmatic and institutional goals. More important, sometimes that egalitarian culture puts faculty members' freedom ahead of the needs of the students. All too often, faculty members want to close their classroom doors and pretend they have their own little fiefdoms. In reality, however, they are part of larger organizations—programs, departments, institutions. Many colleges and universities now see accountability and collaboration as realistic expectations and assessment as an effective means to realize those expectations (Barone, 2003, p. 44). The accreditation process and criteria, such as those outlined in ABET 2000, should help us see that we are indeed accountable and responsible to others. To achieve that accountability to the students and the institution, while balancing the needs of the individual faculty, requires foresight and the involvement of the leadership of the program, department, and perhaps even college. Having a clearer picture of the role of each course in a program, that program's role in the major, and that major's role in the overall university education helps faculty have a larger context for the discussion. By working together to build a program and make the connections within that program explicit, program administrators can assist this process. In the end, individual faculty members will have choices and freedoms, but those choices will include taking larger into account the expectations of others—students, other faculty, administrators, parents, and a range of extended stakeholders.

BENEFITS

In *The Fifth Discipline* (1994), Peter Senge explains, "at the heart of a learning organization is a shift of mind—from seeing ourselves as separate from the world to connected to the world, from seeing problems as caused by someone or something 'out there' to seeing how our own actions create the problems we experience" (pp. 12–13). The shift of mind Senge refers to is sometimes called a "metanoia." As he explains it, "to grasp the meaning of 'metanoia' is to grasp the deeper meaning of 'learning'" (p. 13). For any professional writing program, learning and interconnectedness are central concerns during the program's development; once developed, the challenge administrators face is assessing the success of that learning and the coherence of the interconnectedness. Everyone in the program, from students to faculty, needs to see, understand, and be in accord with the program's objectives, which need to be explicit. The superstructure of the program—how the various courses build upon one and support one another—needs to be visible so that, when tested, the structure will withstand the buffets and blows associated with the winds of political and cultural change and the tensions created by dissensus.

Cargile Cook and Zachry's story about the "metanoia" that occurred at Utah State adds to the scholarship about shaping programs. While I wouldn't say that they displayed "dirty laundry" (p. 66), I would agree that they provided an honest description of the kinds of "turf wars" that arise when faculty work to shape and get to a deeper level of understanding about their program and identity (p. 66). The discussion of the conflict that occurred within their program is a mirror for our field, which is equally conflicted. Those of us in professional writing occupy an uneasy space between theory and practice, which often "revolves around pedagogical decisions related to preparing students to handle the various skills and software they'll need to know to succeed in the workplace, often labeled as an instrumental approach, and preparing them to be reflective, responsible practitioners, who have obligations to their discipline and society, often labeled as a rhetorical one" (Dubinsky, 2004, p. 13). Rather than try to gloss over the tensions, Cargile Cook and Zachry, following the lead of the faculty at Clemson (Yancy et al., 2004), have vividly portrayed the "networks of interests and activities that endlessly complicate our work" (p. 77). The result, which falls in line with what Aristotle would call productive knowledge, is both a making and wisdom about that making.

CONCLUSION

The narrative Cargile Cook and Zachry offer about the complicating factors involved in the less-traveled road to *self*-assessment and program identity formation is an important one; it offers important insights based upon their 7-year journey. Yet, after reading it, I was left with some unanswered questions. I wanted to hear the voices of the curriculum committee during their discussions about the development of a set of ideas and vocabulary with which to define programmatic goals, as they worked to develop self-regulating behaviors, and as they coped with the continual flux of program development (p. 77). I wanted to learn more about the "dissensus," how faculty (or even just one or two as examples) "stamp[ed] students with their own brands of knowledge" and used what they discovered to promote a wide and varied mixture of teaching and learning (p. 84), and whether such a mixture could and did lead to authentic assessment. I wondered how faculty used the students' narratives "about the overall coherence of the program" (p. 69). I wanted to see the way(s) in which a more focused emphasis on ethics changed the program. Hearing that students "consider[ed] all the stakeholders," "describe[d] the . . . research processes," and "critique[d] their documents" (p. 72) was wonderful. Seeing that process in action and understanding how the faculty used these descriptions to effect programmatic change would have helped me better understand the assessment and its value. I also wanted to know more about the scale they asked the students to use concerning literacies. Was the scale discussed in classes? Did faculty all understand its value in similar terms? Did students? How were the students "taught" to judge the various issues involved and compare them with what they knew before? I very much wanted to understand more about the shared knowledge and what was agreed upon by the faculty as a whole. I was also curious whether or not there were those in dissent and what they did, if anything, to implement or undermine the objectives.

Finally, I wanted to more fully understand the model for program assessment that Cargile Cook and Zachry presented: integrating portfolio assessment into a capstone course. Portfolios have been gaining ground since the mid-1980s, when composition teachers such as Peter Elbow and Pat Belanoff adopted them as an alternative assessment strategy. In the 1990s, they made another breakthrough following the Miami University Conference on Portfolios (1992), and more recently, with the advent of increasingly available open-source software, universities are adopting electronic portfolios, which some scholars are claiming "have a greater potential to alter higher education at its very core than any other technology application we've known so far" (Batson, 2002, para. 4). For example, some schools, such as the Rose-Hulman Institute of Technology, with its RosE-Portfolio, focus on "specific learning objectives in areas such as ethics, contemporary issues, culture, and communication" (Greenberg, 2004, p. 32). At Virginia Tech and other schools, ePortfolios are being tested to see how they might work as assessment vehicles, as well as ways to showcase student work and offer students opportunities to reflect on their learning processes. If they can indeed serve these important functions, faculty may have to rethink ways to collaborate, communicate, and negotiate program identity and learning communities.

All those musings offered, I conclude by saying that those questions are not flaws in the essay or the work our colleagues have completed. Rather, they are a fruitful result of a provocative and important story that Cargile Cook and Zachry shared, which I hope will be just one of many installments that will come from other program administrators in our field. The benefits from such stories that demonstrate the importance of reflection and programmatic self-assessment to create a culture of collaboration leading to change cannot be underestimated. As I reflect further on the journey these authors and their colleagues took, I realize, without question, that some or all of the insights they shared will help my colleagues and me more accurately outline a roadmap for assessment in our program. Because of their travelogue, we might be able to spot better stops along the way and avoid some of the difficult stretches of road.

REFERENCES

Argyris, C., & Schön, D. (1975). *Theory and practice: Increasing professional effectiveness*. London: Jossey Bass.

Batson, T. (2002, December). The electronic portfolio boom: What's it all about?" *Syllabus, 16*(5). Retrieved March 10, 2003, from http://www.syllabus.com/article.asp?id=6984

Barone, C. A. (2003, September–October). The changing landscape and the new academy. *Educause, 38*(5), 40–46.

Cargile Cook, K., & Zachry, M. (2009). Politics, programmatic self-assessment, and the challenge of cultural change. In M. N. Hundleby & J. Allen (Eds.), *Assessment in technical and professional communication* (pp. 65-79). Amityville, NY: Baywood

Dubinsky, J. M. (2004). Introduction to chapter 1: Introducing theoretical approaches. In J. M. Dubinsky (Ed.), *Teaching technical communication: Critical issues for the classroom*. Boston, MA: Bedford/St. Martin's.

Greenberg, G. (2004, July–August). The Digital Convergence: Extending the portfolio model. *Educause, 39*(4), 28–36.

Polanyi, M. (1967). *The tacit dimension*. Garden City, NY: Doubleday.

Senge, P. (1994). *The fifth discipline*. New York: Currency Doubleday.

White, E. M. (1998). *Developing successful college writing programs*. Portland, ME: Calendar Island Publishers.

Yancey, K., Williams, S., Heifferon, B., Hilligoss, S., Howard, T., Jacobi, M., et al. (2004). Notes toward 'reflective instrumentalism': A collaborative look at curricular revision in Clemson university's MAPC program. In T. Bridgeford, K. S. Kitalong, & D. Selfe (Eds.), *Innovative approaches to teaching technical communication* (pp. 93–108). Logan, UT: Utah State University Press.

Situating Assessment in Disciplinary Requirements

Expanding the Role of Technical Communication Through Assessment: A Case Presentation of ABET Assessment

Michael Carter
North Carolina State University

Technical communication teachers should not ignore the most extensive venue for assessment of technical writing on many of our campuses, assessment by outside, disciplinary, or professional accreditation agencies. The best-known example is the Accreditation Board of Engineering and Technology (ABET), whose current accreditation procedure, launched as Engineering Criteria 2000, invites engineering programs to create their own standards of assessment. A radical departure from previous ABET "bean counting" that stressed consistency across all accredited engineering programs, the new emphasis on local standards and the continuous and ongoing assessment of programs according to those standards offers technical communication teachers an opportunity to make a critical contribution to the role of writing and speaking in engineering majors— and suggests the possibility for taking advantage of other sites of assessment on our campuses.

This chapter considers the case of one engineering program and my participation in the assessment process of that program, which became a model for other programs in the College of Engineering at my university. In 1997 North Carolina State approved a plan for the outcomes-based assessment of writing and speaking in undergraduate majors. I was responsible for helping faculty in departments identify writing and speaking outcomes appropriate for their majors and for creating and implementing plans for assessing those outcomes. In the 2nd year of this process, I was scheduled to work with the College of

Engineering. The associate dean of the college recognized the correlation between what I was doing and the then-new and untried ABET assessment guidelines calling for outcomes-based assessment. I agreed to take what I had learned working with other departments and broaden the scope to address ABET guidelines. We identified four pilot departments for our initial effort, one of which was civil engineering (CE), the subject of this case presentation. I consulted with the college for the next 2 years on ABET assessment and, on the basis of the pilots, created extensive materials to guide the assessment activities of other programs in the college (Carter, 2001a, 2001b).

ABET accreditation consists of eight general criteria that programs are required to address. I will focus here on Criterion 2, Program Educational Objectives, and Criterion 3, Program Outcomes and Assessment (ABET, 2004c), the two criteria that have elicited the most consternation among engineering faculty preparing for ABET reviews and that offer the best opportunities for technical communication teachers to participate in engineering program assessment. The other criteria, which include issues such as faculty, finances, and facilities, are relatively straightforward, similar to the previous ABET model. It is in Criteria 2 and 3, however, that engineering faculty must establish the standards and create plans for assessing their programs. And it is here that writing and speaking can play an important role.

Generally speaking, there are three levels of ABET participation possible for technical communication teachers. One is to focus only on the communication outcome, Criterion 3g, "an ability to communicate effectively" (e.g., Hovde, 1999, 2003b). Engineering faculty may need help in defining in operational terms what it means for students to write effectively and how to measure that ability. Also, the assessment of students' communication skills could include their work in technical communication courses. Teachers of those courses could take advantage of such an opportunity to create stronger ties to engineering programs and to play a key role in the process of designing and implementing assessment.

A more expansive level of participation is to take advantage of the potential for writing and speaking in other outcomes from Criterion 3 and also the objectives of Criterion 2. ABET's requirement for direct evidence for assessment means that there may be other venues in which students' writing and speaking could be used as evidence for outcomes assessment. For example, Criterion 3c, "an ability to design a system, component, or process to meet desired needs," could be assessed with students' technical reports and oral presentations from capstone design courses. Using writing and speaking to evaluate some of the other general outcomes, such as 3f, "an understanding of professional and ethical responsibility," may demand more imagination but also offers the opportunity to incorporate writing and speaking in meaningful ways in the curricula of engineering students. On these first two levels, technical communication teachers can provide much-needed assistance to their engineering counterparts and, in the process, help them to become more aware of writing and speaking as both tools for learning and artifacts for assessing learning.

The broadest level of participation is to play an integral role in creating, assessing, and documenting the objectives and outcomes of Criteria 2 and 3 for ABET review. I believe that experts in technical communication are ideally situated to act as consultants and even to provide leadership in the broader ABET process: we understand the community and discourses of engineering and can provide valuable guidance to engineering faculty as they interpret and assess the ABET criteria. It is this last level that is the primary focus of this case presentation, though I also address the opportunities to contribute specifically to writing and speaking appropriate to the other levels. Whatever the extent of engagement in the ABET process, the result could be a productive conversation between engineering faculty and technical writing teachers that could lead to a much better understanding on the part of everyone concerned about the place of writing and speaking in our students' broader curriculum and the way technical communication courses fit into that curriculum.

The purpose of this chapter, then, is to highlight the opportunities that accreditation offers to technical communication teachers for participating in assessment on their campuses, opportunities we may not be aware of as we focus on our own courses and programs. Though the emphasis in this case is on ABET, the procedures described here may be generally applied to working with programs under the aegis of other professional accreditation agencies requiring outcomes-based assessment, including such diverse programs as social work (Council on Social Work Education), business (Association to Advance Collegiate Schools of Business), and education (National Council for Accreditation of Teacher Education). Technical communication teachers are well positioned to help faculty across our colleges and universities use writing and speaking effectively for meeting academic standards. But the implications extend even more broadly than that. Regional accreditation agencies, such as the Southern Association of Colleges and Schools,[1] are now requiring outcomes-based assessment for all college and university programs, which means that academic programs across our institutions may find a fresh interest in their students' writing and speaking, not only as tools for teaching but also as evidence of student learning. In addition, we technical communication teachers are finding ourselves in the position of assessing our own technical communication and professional writing programs. Outcomes assessment on all of these levels provides the potential for a broader, institutional improvement of our students' writing and speaking.

---

[1] For example, the Southern Association of Colleges and Schools says in its guidelines, "The institution identifies expected outcomes for its educational programs and its administrative and educational support services; assesses whether it achieves these outcomes; and provides evidence of improvement based on analysis of those results" (Commission on Colleges, 2004, p. 22).

CREATING PROGRAM EDUCATIONAL OBJECTIVES: ABET CRITERION 2

Program Educational Objectives (PEO) are arguably the most important part of the ABET criteria. PEO embody the vision for an engineering program, a vision that projects the goals for the broader assessment process. Criterion 2 is the least defined of all the ABET criteria. Unlike Criterion 3, for example, which provides a list of curricular outcomes that should be met, Criterion 2 offers little in the way of direct guidelines, neither for what should be included in the PEO nor for the process of generating and assessing them:

> Each engineering program for which an institution seeks accreditation or reaccreditation must have in place:
> (a) detailed published educational objectives that are consistent with the mission of the institution and these criteria
> (b) a process based on the needs of the program's various constituencies in which the objectives are determined and periodically evaluated
> (c) a curriculum and processes that prepare students for the achievement of these objectives
> (d) a system of ongoing evaluation that demonstrates achievement of these objectives and uses the results to improve the effectiveness of the program (ABET, 2004c, p. 1).

Though these instructions describe in general terms what each program should accomplish, they don't indicate how they should be accomplished. Neither do they say what documents should be presented to the ABET reviewers for evaluating the degree to which Criterion 2 has been satisfied. In the case presented here, I describe three tasks for helping faculty in the CE program manage Criterion 2, the development of (1) criteria for good PEO and for an effective procedure for establishing and assessing PEO, (2) a procedure for establishing and assessing PEO, and (3) a list of documents related to PEO to be provided to ABET reviewers.

Criteria for Good PEO and for an Effective Procedure for Establishing and Assessing PEO

What are PEO? What procedure should be used for establishing and assessing PEO? How do you know if PEO are sufficient and appropriate? Engineering faculty ask these questions but find few answers in the ABET instructions. The problem is that the instructions lay out the expectations for what should be in place for the review but give little guidance in how programs are to achieve those expectations. What is needed, then, is a set of parameters to limit and guide the work on PEO. To create such parameters, I searched the information about Criterion 2 provided by ABET and pulled out implicit and explicit criteria for good PEO that could provide a workable goal structure for CE faculty.

A key source of information was the Self-Study Questionnaire, published by ABET (2004a). Particularly helpful was the "Matrix for Implementation Assessment," a rubric included in the questionnaire. For instance, the very best PEO are described as "Comprehensive; defined[,] documented, measurable and flexible; clearly tied to mission; readily adaptable to meet constituent needs; systematically reviewed and updated." Even these explicit criteria needed to be unpacked to be useful to faculty. So I used the instructions and the matrix to create a set of elaborated criteria for PEO:

Program Educational Objectives should be

- **consistent with the mission of the institution** (they should reflect pertinent parts of the mission statements of the university and the College of Engineering and the department)
- **consistent with the needs of key constituencies** (they should address the concerns of the major stakeholders in the program, such as students, employers, and industry advisory boards)
- **consistent with the rest of ABET criteria** (they should include, where appropriate, issues related to Criterion 1 and Criteria 3–8)
- **comprehensive** (they should be broad and overarching, providing a vision for the whole program)
- **defined** (they should be clearly delineated with enough detail to make them meaningful for the program)
- **measurable** (they should be written in a way that allows for qualitative and quantitative assessment)
- **flexible** (they should not narrow or unduly limit the possibilities of the program but should be adaptable to changes in the needs of constituencies and mission of the institution; in other words, they should incorporate the language of vision rather than the language of long-range planning)
- **published** (they should be made public, for example, placed on the program's Web site, included in brochures about the program, printed in newsletters sent to alumni, given to students upon entry to the program) (Carter, 2001a).

In addition to the criteria for sufficient and appropriate PEO, it was also necessary to elaborate criteria for an effective procedure for establishing and assessing PEO. One of the crucial goals of ABET accreditation is to encourage continuous and ongoing assessment of engineering programs; thus, there must be a procedure created and implemented for guiding programs in attaining that goal. Faculty, then, need to have criteria for an effective procedure in order to create one that allows them to meet the high standards of ABET reviewers. Using the PEO instructions and criteria from the matrix, I listed criteria and elaborated on them in the following way:

An effective process for establishing and assessing PEO should

- **provide for a high degree of ongoing involvement of constituencies of the program** (those that are identified as key constituencies of the program should play an important role in defining and assessing PEO and in the ongoing cycle of improvement in the achievement of PEO)
- **attend to the mission of the institution** (faculty need to find the mission statements and associated documents for the university, the College of Engineering, and the department and read them carefully to identify any specific language or general ideas that should be included in the PEO; the purpose is to demonstrate through the PEO how the department contributes to enabling the college and university to achieve their missions)
- **delineate a clear methodology for assessing PEO** (assessment must set forth for each PEO the data to be collected and how the data will be analyzed)
- **establish a plan for assessing the PEO** (it is necessary to have assessment cycles that occur with enough frequency to give programs a good sense of the extent to which they are achieving their objectives and to act on what they find for the improvement of the program: when data are to be collected and analyzed, who is responsible for collecting and analyzing data)
- **provide the means for applying the results of assessment directly to improving the program** (assessment of PEO should establish who is responsible for evaluating the results and for applying what has been learned from results to improvement of program) (Carter, 2001a).

A Procedure for Establishing and Assessing PEO

The criteria listed above make more explicit what faculty should aim for in establishing and assessing PEO, thus implying a procedure for achieving those two tasks. The next step, then, was to make that procedure explicit, which entailed outlining a process that would enable faculty to meet the process criteria.

The specific tasks that need to be performed for establishing and assessing PEO:

1. **List the key constituencies of the program.** Constituencies are the various categories of people who have some vital interest in your program because they are in some way affected by it (students, faculty, employers, recent graduates, parents of students, etc.).
2. **Identify the individuals or groups you will use as representatives of those constituencies.** Most constituencies are too large to be used as a whole, so you will need to identify suitable representatives for the groups and/or describe methods, such as a survey, for gaining the participation of a large group.

3. **Create a step-by-step plan for establishing PEO.** Detail the process you intend to follow, including proposed dates and times for various meetings, deadlines for gathering various kinds of information, the parties responsible for getting tasks done, and the key constituencies.

4. **Establish a final list of PEO based on a variety of inputs.** The PEO represent a vision for the program, the goals it sets for itself to achieve, clearly tied to the mission of the institution—university, college, and department—and should endeavor to meet constituent needs.

5. **Describe the relationship between the PEO and the needs of key constituencies and the institutional mission.** A crucial aspect of Criterion 2 is to show that through its PEO, the program is responding to the needs of the key constituencies and serving to further the missions of the university, the college, and the department. Using either list or graphic form, describe these links.

6. **Describe the relationship between the PEO and the program curriculum.** Diagram the connections between the objectives and the curriculum, as represented by the curricular outcomes defined under Criterion 3 and the present courses in the curriculum.

7. **Describe the relationship between the PEO and the other EC 2000 criteria.** This task helps you to start thinking seriously about how you will assess the PEO. You can use much of the data you collect for these other criteria in the assessment of the PEO.

8. **Outline a procedure for periodic assessment of PEO that**

   (a) establishes review cycles,

   (b) describes the various kinds of data that will be used for reviews (probably a mix of qualitative and quantitative data, which may include data from assessments of the other criteria),

   (c) outlines a detailed review process that will include key constituencies (you may want to use a revised version of the procedure for establishing the objectives), and

   (d) presents a clear method for using the data for continual improvement of the program and for updating the objectives in response to changes in the particular field of engineering, the needs of the constituencies, and the mission of the institution.

9. **Report results for each of the review cycles by**

   (a) chronicling the review process,

   (b) presenting the primary data gathered for the process,

   (c) describing the evaluative outcomes of the review in terms of each of the objectives, and

   (d) showing how the outcomes of the review were incorporated into the larger process of the continual improvement of the program and/or led to an updating of the objectives (Carter, 2001a).

Documents Related to Criterion 2 to be Provided for ABET Reviewers

Giving engineering faculty a better understanding of the goals they should be aiming toward in Criterion 2 and how they may achieve the goals provides them a much-needed sense of direction. For most faculty, however, the bottom line in this entire process is what they need to produce for ABET reviewers. The instructions for Criterion 2 and the details of the self-study report from the questionnaire provide a good idea of what evaluators will expect to see when they visit the campus. But these guidelines are still quite vague.

The faculty I worked with wanted a much more specific description of the kinds of documents they should produce for the ABET review, documents that would give reviewers the information they needed to rate the program in terms of the matrix. The reviewers would need to see evidence of continuous and ongoing improvement, and that evidence must be organized clearly enough that reviewers could find what they needed efficiently.

The idea, then, was to take the implicit guidelines presented by ABET and create a set of documents that would reflect those guidelines and meet the needs of the ABET reviewers. The list of documents I developed for CE was designed to help faculty to meet both of those standards.

Documents to be provided to ABET reviewers

1. **"Program Educational Objectives,"** stating the objectives and then providing supporting information, including
   (a) a list of the PEO;
   (b) a detailed discussion of the PEO, showing how each one is consistent, where appropriate, with the mission of the institution, the needs of key constituencies, the requirements (if any) of particular professional accrediting agencies for the program, and with the spirit of continual improvement represented in the ABET criteria;
   (c) a description of the relationship between the objectives and the program's curriculum (via Criterion 3) and other relevant ABET criteria.
2. **"Establishing Program Educational Objectives,"** describing your process for establishing the PEO, emphasizing the degree of involvement of the key constituencies and the role of the institutional mission in the process. In this document you ought to include
   (a) a list the key program constituencies,
   (b) a discussion of how and why those constituencies were identified as key to the process of establishing objectives,
   (c) a delineation of the way individuals or groups were chosen to represent the key constituencies; how and why they were selected,
   (d) a description of the initial plan for establishing the objectives and how you arrived at that plan, and
   (e) a detailed outline of the process you followed.

3. **"A Process for Periodic Review of Educational Program Objectives,"** showing a thorough and realistic procedure that allows you to review, in consultation with key constituencies, the extent to which you are achieving the objectives and also to review the objectives themselves, all of which leads to a more effective and responsive program. This document should include

   (a) a formal outline of the review procedure: assessment cycles, kinds of data to be used for evaluation, how key constituencies were involved in the review, how the results of the review were used to improve the effectiveness of the program and/or revise the objectives;

   (b) reports for each of the review cycles detailing the particular process undertaken, the results of the review, the recommendations for any changes in the program or changes in the PEO, and the process of implementing the changes (Carter, 2001a).

The critical step for Criterion 2 is the first step, establishing the criteria for good PEO. Developing these criteria requires an interpretation of the ABET documents, extracting and elaborating on the implicit and explicit standards embedded in these documents. These criteria make Criterion 2 more concrete, giving engineering faculty a clearer direction to follow while also providing a foundation for a procedure for establishing and assessing PEO and for documents to be produced for ABET reviewers (see Appendix A for PEO produced by CE faculty based on the above materials.)

Criterion 2 also offers opportunities for technical communication teachers who prefer to focus specifically on writing and speaking. The most recent ABET guidelines, perhaps in response to complaints that Criterion 2 was too vague, say that "program educational objectives are intended to be statements that describe the expected accomplishments of graduates during the first several years following graduation from the program" (ABET, 2004c). Certainly those expected accomplishments should include graduates' communication abilities. Technical communication teachers could work with engineering faculty to write objectives that adequately describe the expectations, to design appropriate assessment methods, to interpret assessment results, to advise faculty in improving writing instruction as indicated by the results, and to report assessment activities for reviewers (Hovde, 2003a). The results of such assessment procedures could also be of great value in improving technical communication courses for engineering majors.

CREATING PROGRAM OUTCOMES: ABET CRITERION 3

Criterion 3 provides engineering faculty firmer footing than Criterion 2 because the former is concerned with curricular issues, what happens in the classroom. Also, Criterion 3 provides a list of general program outcomes to be assessed, giving engineering faculty a clearer set of parameters. These 13 outcomes, as anyone even a little familiar with ABET accreditation will know, are listed as *a* through *k*, ranging from "an ability to apply knowledge of mathematics, science, and engineering" to "a knowledge of contemporary issues." ABET has made it clear that engineering programs are expected to provide educational opportunities related to all 13 outcomes, and that all students in the program are expected to attain all outcomes (ABET, 2004b).

Though the parameters of Criterion 3 are somewhat clearer than Criterion 2, Julia M. Williams (2001) points out that they are still "annoyingly vague" for engineering faculty. Technical communication teachers have the ability to help their engineering colleagues meet the three main challenges of Criterion 3: (1) to define each of the general outcomes so that they are both appropriate to the program and are measurable; (2) to establish a viable plan for assessing the outcomes; and (3) to provide adequate documentation to ABET reviewers, describing the process used to accomplish the first two tasks and reporting the assessment results and how those results were used to improve the program. I will discuss an approach for addressing each of these challenges. For a description of a procedure I developed for determining educational outcomes and plans for assessing those outcomes, see Carter (2002).

Defining Program Outcomes

One of the goals of ABET's outcomes-based assessment is to encourage engineering faculty to define their own programs: a radical departure, as I mentioned earlier, from the former cookie-cutter approach to accreditation. Criterion 3 plays a critical role in the process of defining a program. Instead of prescribing the kinds of courses all engineering students should take, ABET now allows each program to establish an outcome for itself and then shape its curriculum so that students can achieve the outcome.

Let's take, for example, Criterion 3b: "an ability to design and conduct experiments, as well as to analyze and interpret data." The four infinitives describe in general terms what all engineering students are expected to learn to do. The faculty of each program, then, are supposed to operationally define what it means to accomplish those activities for their program. In addition, they are supposed to write that definition in a way that is both teachable and measurable; that is, the definition should provide enough specificity to guide teachers as they address the stated abilities in their classes or other educational venues and to allow for meaningful measurement of the abilities.

Thus, the challenge for engineering faculty is to define the general statements listed in Criterion 3 in a way that is appropriate to their programs and sufficiently detailed to be teachable and measurable. In my work with CE faculty, I realized that starting with a blank slate for each of the outcomes was daunting, so I created a template: a set of broadly stated outcomes that would serve as a starting point for discussion. The template for 3b was

To demonstrate that graduates have an ability to design and conduct experiments as well as analyze and interpret data, they should be able to

- take an experimental problem and develop a hypothesis, define the pertinent dependent and independent variables, and establish a sound experimental method that will allow them to measure the variables and test the hypothesis;
- conduct an experimental procedure, use laboratory materials properly and safely, carefully note observations in a laboratory notebook, and describe the procedure clearly for others;
- measure and record raw experimental data and analyze those data for the purposes of understanding and explaining the data. Graduates should be able to represent data in both verbal and visual forms (equations, tables, graphs, figures, etc.) in a way that is both an accurate and an honest reflection of the data.
- render the data meaningful by discussing the data in the context of the hypothesis and appropriate theories and principles and by stating, clearly and concisely, conclusions that can be drawn from the experiment (Carter, 2001b).

This template is based roughly on the parts of the standard laboratory report.

My strategy with the CE faculty committee I was working with was to present the template to them and allow them to use it as a starting point for making an operational definition of the outcome that better reflected their own program. The faculty immediately noted that the language and some of the concepts did not really reflect their sense of civil engineering. For example, they felt uncomfortable with the portrayal of civil engineers as laboratory scientists and added "field tests" to give a better sense of what they expect their student to do, even to the point of rewriting the ABET outcome. But they also realized that they couldn't make the language of the outcome too specific because their program included, in addition to standard civil engineering, construction and environmental engineering. The final version of the outcome, then, found a balance between the general and the specific and between what civil engineers do and what ABET wants.

To demonstrate that graduates have an ability to design and conduct civil engineering laboratory and field tests as well as analyze, interpret, and present data, they should be able to

- design civil engineering laboratory and field tests in terms of collecting and measuring samples, knowing where to take measurements, deciding how many measurements to take, and demonstrating an understanding of accuracy and precision.
- conduct laboratory or field tests properly and safely and describe the procedures they used to conduct those tests so that others can understand them.

- measure and record raw test data and analyze those data for the purposes of understanding and explaining the data. Graduates should be able to interpret and present data in both oral and visual forms (equations, tables, graphs, figures, etc.) in a way that is an accurate reflection of the data.

- use data to make sound judgments in the solution of problems in civil engineering and support those judgments and solutions according to standards of practice in the field (Carter, 2001b).

The result is an operational definition that is appropriate to this particular civil engineering program and is also teachable and measurable. It spells out what students should be able to do in a way that gives lab instructors and professors guidance in their teaching and in a way that renders ABET's general program outcome in sufficient detail that it can be assessed by the faculty.

I found that the template method also worked well for other programs. It provided engineering faculty a model of what an operational definition of an outcome could look like but also allowed them to recognize where their programs were not reflected in the template and revise it accordingly.

Establishing a Plan for Assessing Outcomes

ABET's "Criteria for Accrediting Engineering Programs" (2004) says, "Each program must have an assessment process with documented results. Evidence must be given that the results are applied to the further development of the program. The assessment process must demonstrate that the outcomes of the program . . . are being measured" (p. 2). Measurement is central to ABET evaluation of programs. Criterion 3 requires that programs create a process that allows them to determine and document how well students are able to meet the stated program outcomes, a through k and any others that have been stated. However, because the focus of engineering faculty tends to be on measuring outcomes, when we are generating outcomes, I try to keep the conversation centered on faculty expectations of what students should be able to do. Once the outcomes have been identified, faculty can turn their attention to creating an assessment plan.

An assessment plan identifies evidence to be gathered for each outcome and establishes a timeline for when the data are to be gathered and analyzed. ABET specifies that that the "assessment process should include direct and indirect measures and does not rely only on self-report surveys and evidence that is 'covered' in the curriculum" (ABET, 2004b, p. 2). Thus, programs cannot simply point to courses taken by students and to surveys of students as evidence. There must also be more direct measures, such as "student portfolios; subject content examinations; performance evaluation of work/study, intern or co-ops; and/or performance observations" (p. 3). One of the challenges, then, is helping faculty identify direct forms of measurement that can be complemented by indirect measures.

In the case I am following here, I will continue my focus on Criterion 3b. Faculty on the CE committee decided on four kinds of evidence to be used to evaluate how well their students could design, conduct, and manage the data of laboratory experiments and field studies:

- Course syllabi [for evidence that items related to effective lab experimentation have been covered in classes]
- Samples of student work (laboratory reports from courses such as CE 332, 324, 381, 342)
- Graduating Senior Survey (questions #2, #10, and #13) [students' self-report on appropriate issues]
- Faculty Perception Survey [faculty impressions of students' knowledge and abilities associated with 3b] (Carter, 2001b)

The primary evidence of Criterion 3b is students' lab reports, direct evidence of the four elements of that outcome as operationally defined by the faculty. The rest of the evidence is indirect and provides secondary information about the outcome.

Documenting the Process of Assessing Criterion 3

ABET's Self-Study Questionnaire (2004a) provides a useful description of the information that programs are expected to provide ABET reviewers. Using that as a guide, I created a list of documents faculty could prepare for the review:

Documents to be provided to ABET reviewers:

1. **"Program Outcomes,"** demonstrating how the outcomes have been operationally defined for the program and providing information related to the outcomes, including
   (a) a list the operational definitions of outcomes, a through k and any others that have been added;
   (b) a description of how particular outcomes relate to specific Program Educational Objectives;
   (c) a description of how each of the outcomes relates to the general outcomes a–k in Criterion 3;
   (d) a plan for assessing outcomes, including the evidence to be collected for each outcome and a timetable for collecting and analyzing the evidence;
   (e) a description of the process used to generate the outcomes and the assessment plan.
2. **"Report of Outcomes Assessment,"** presenting the findings of your assessment. If you have more than one assessment cycle, report the results for each cycle. You may think of it as the methods, results, and discussion sections of a paper and should include

(a) a description of the assessment methods you used for each of the outcomes, methods for both collecting and analyzing the data (these methods may be different in application from the assessment plan);

(b) a summary of the findings of the measures for each of the outcomes;

(c) a discussion of the conclusions of the findings in terms of students' satisfactory attainment of each outcome;

(d) a description of any changes in the program based on the findings of the assessment, specifically referring to the data used as a basis for the changes;

(e) a description of the materials (including both direct and indirect evidence organized by outcome) that will be made available during the ABET review (Carter, 2001b).

EVALUATING WRITING AND SPEAKING FOR ABET ASSESSMENT

Once faculty have generated objectives and outcomes and plans for assessing each, then they can begin the process of gathering and analyzing data. ABET's emphasis on direct evidence means that this evidence will likely include samples of students' writing and speaking. Engineering faculty assessing these data need help in evaluating written and oral performance of their students and in drawing meaningful conclusions from the evaluation. Perhaps no one on campus is better qualified to provide such help than technical communication teachers. I will describe my work with CE faculty to guide their evaluation of students' lab reports for outcome 3b. This experience allowed me to create a general procedure that other programs in our College of Engineering could use to create and implement evaluation rubrics.

The ABET committee of the Department of Civil Engineering had identified student lab reports as direct evidence for Criterion 3b. They turned this assessment over to the Laboratory Committee, composed of faculty responsible for the various CE labs. The first task of the Laboratory Committee was to determine what procedures it would use for collecting and analyzing data. In order to include students from the three different fields within the CE department, the committee decided to collect lab reports from all four of the labs listed as possibilities on the assessment plan, enabling them to cover all majors. They would ask the professors responsible for those labs to copy the reports that they thought best represented the full scope of the abilities defined in the program outcome. Random samples of the reports from each course would be evaluated by members of the Laboratory Committee at the end of the designated semester.

Once the logistics of evaluation had been established, the committee turned its attention to creating an evaluation rubric. We began with a careful review of the stated outcomes for Criterion 3b, looking for the implicit criteria for the rubric. The first draft of it consisted of criterion statements straight from the outcomes (see Appendix B). Over the next two meetings, the committee restructured, elaborated, and amended those initial criteria to create a rubric that they believed more clearly specified the textual features that marked the abilities defined in the outcome.

Next, the committee piloted the rubric that resulted from their revisions by applying it to sample reports that some of the members brought from their labs. This pilot evaluation was valuable because it enabled the members of the committee to uncover several problems in the rubric, the lack of clarity in some of the criteria, some redundancy among the criteria, and the terms used to evaluate each criterion. As to the last of these, the original evaluative terms on the rubric were "poor, fair, excellent." After the pilot, the faculty realized that they needed to include a "not applicable" to designate elements of the reports that were missing or were not applicable to the particular kind of report. They also felt uncomfortable scoring any of their colleagues' student reports as "poor," so they decided to use "needs improvement" instead and to elevate "fair" to "good."

The pilot not only provided valuable feedback for improving the rubric but it also demonstrated to the members of the committee that they could indeed use the rubric and that the rubric worked—it enabled them to make useful evaluative judgments about students' abilities to apply scientific reasoning to labs and field tests.

At the end of the semester, the committee formally evaluated lab reports from the four lab courses. The final version of the rubric (see Appendix C) followed the outcome by structuring the criteria according to the main abilities defined in it: (1) "to design civil engineering laboratory or field tests," (2) "to conduct civil engineering laboratory or field tests properly and safely," and (3) "to analyze and interpret data from civil engineering laboratory and field tests." Faculty were satisfied that the rubric successfully translated the implicit criteria in the outcome into textual features of lab reports. After the evaluation, committee members discussed their findings, agreeing, for example, that the first three criteria, which were included in the introduction, did not meet their expectations. At the next meeting, the committee was given a spreadsheet containing the evaluation results and discussed problems and possible solutions. The resulting report, drafted by the committee chair, gave the data from the evaluation, identified shortcomings that existed across samples from all four courses, and made recommendations for addressing these shortcomings in the labs.

I generalized the procedure developed for CE into six steps that can be followed by other programs:

1. Identify data (i.e., type of student performance) appropriate for assessing the outcome (see outcome and assessment plan).

2. Determine logistical procedures for collecting and assessing data.
   - specific data to be collected (which papers or presentations are to be collected from which classes)

- when the data are to be collected
- sampling technique to be used (if appropriate)
- who is responsible for collecting and copying data
- who will perform the assessment
- when and how often assessment will take place

3. Generate an assessment rubric (set of criteria) **based on the outcome to be assessed.**
   - review outcome to identify basic criteria
   - elaborate on basic criteria where necessary (establishing specific criteria appropriate to the kind of student performance)
   - revise criteria so that they describe what students will be doing in the writing or speaking performance that will be evaluated
   - create a preliminary list of criteria for discussion and further revision and eventual approval
   - turn approved list into an evaluation rubric (a grid with criteria and evaluative terms, such as "poor, fair, excellent") for each criterion

4. Pilot rubric with appropriate evaluators in order to identify any problems with it; revise accordingly.
   - use actual evaluators or a similar group
   - use actual or similar samples of student performance
   - ask the evaluators to identify any problems with using the rubric: misunderstandings of criteria, criteria that don't seem to be relevant, criteria that are missing, etc.
   - revise rubric

5. Apply rubric in real assessment of student performance, including, where appropriate, proper training for evaluators.
   - go over criteria and answer questions about them
   - present one or two evaluated samples
   - ask evaluators to assess one or two common samples and discuss any differences that arise, establishing basic agreement among evaluators

6. Ask evaluators to discuss the results of the assessment and come to a collective judgment as to the degree to which the program is enabling students to achieve the outcome; write and submit assessment report to appropriate audience.
   - describe procedure used for evaluation
   - describe results (quantitative and/or qualitative)
   - outline evaluators' judgment of strengths and areas for improvement
   - recommend areas the program can address for improvement (Carter, 2001b).

I stressed to the engineers I worked with that they should not feel compelled to achieve the standards of publishable educational research, unless, of course, they intended to publish their work. As soon as the members of the Laboratory Committee had developed an evaluation procedure, they began to search for statistically sound sampling techniques, to fret about interrater reliability, and to voice doubts that they could ever make the procedure work. I reminded them that the purpose of the assessment was to provide evidence for making a judgment about the extent to which the program prepares students to meet the laboratory outcome, and that a judgment doesn't require rigorously applied sampling techniques and a high interrater reliability. It requires that faculty read enough lab reports to satisfy themselves that they are able to discern students' strengths and weaknesses according to the criteria. They frequently make judgments about their courses and curricula. The advantage in this case is that their judgment is based on evidence that has been systematically gathered and evaluated.

Within this framework of assessment, validity is largely a function of the faculty's faith in the rubric to measure what it should measure and thus to guide a reasonable judgment about the outcome. A procedure like the one described above, which fully involves faculty in creating and testing the rubric, is critical to the faith the faculty have in the evaluation. I suspect that a rubric provided by an outsider would not generate the same faith in its validity.

CONCLUSION

Technical communication teachers should take advantage of the extensive and continuous assessment instituted by ABET to seek out opportunities for improving students' writing and speaking. Our participation in that assessment can range from mining assessment data for information that could have implications for our courses to providing leadership in the full assessment process. I have described the latter role here but have also included discussions of more focused roles.

The benefits of participating in some way in engineering assessment could be substantial. Hovde, for example, describes benefits for both technical communication and engineering faculty. For the former, she includes the potential for changing our own courses and curricula based on our interaction with engineering faculty and the opportunity to demonstrate how important communication courses are for engineering majors. For the latter, she includes the development of knowledge about writing that allows them to articulate their expectations for their students' communication and elevating the status of communication in engineering programs (Horde, 1999). ABET assessment is an enterprise that invites engineering faculty to consider all the components of a curriculum and how each contributes to the development of critical abilities of students in their programs and the performance of graduates in their first years on the job. We need to be a part of the broad conversation that has as its goal the improvement of our students' education, an essential component of which is their writing and speaking.

The case I have presented here has implications beyond engineering programs (see also Williams, 2001). Professional accreditation in other disciplines and regional accreditation have extended outcomes-based assessment to academic programs across our colleges and universities. As a result, we technical communication teachers could be presented with an opportunity for a much wider conversation about writing and speaking on our campuses. We should be at the center of that conversation, a position that could also result in a leadership role in assessment efforts on our campuses. Mostly, we can take advantage of the opportunities offered by assessment to integrate our courses effectively within the institutional network and to help other faculty across the campus to better understand the important role writing and speaking play in the education of our students.

APPENDIX A:
Program Educational Objectives

Program Educational Objectives
Civil Engineering
North Carolina State University

In collaboration with representatives of the significant constituency groups, the faculty of the department developed the following educational objectives for the program in Civil Engineering. These objectives are published in the undergraduate catalog and appear on the department's Web site.

1. To prepare students for entry into successful careers in civil engineering, emphasizing the mastery of engineering fundamentals, the ability to solve engineering problems, the importance of engineering judgment and engineering experimentation, and the process of engineering design.
2. To instill in students the sense of pride and confidence that comes from applying their knowledge of engineering principles and procedures to the economic and social benefit of society.
3. To encourage in students an understanding of the professional and ethical obligations of the engineer, to conduct themselves as professionals, recognizing their responsibility to protect the health and welfare of the public, and to be accountable for the social and environmental impact of their engineering practice.
4. To establish an educational environment in which students participate in multidisciplinary, team-oriented, open-ended activities that prepare them to work in integrated engineering teams.
5. To offer a curriculum that encourages students to become broadly educated engineers and life-long learners, with a solid background in the basic sciences and mathematics; an understanding and appreciation of the arts,

humanities, and social sciences; an ability to communicate effectively for various audiences and purposes; and a desire to seek out further educational opportunities.

6. To expose students to advances in engineering practice and research as preparation for opportunities in professional practice and graduate education.

7. To acquire, maintain, and operate facilities and laboratory equipment appropriate to the civil engineering program, and to incorporate traditional and state-of-the-art technology and methods.

8. To recruit, develop, and retain faculty who are committed to the educational mission of the civil engineering program, to ensure that these educational objectives are met.

How these Program Educational Objectives fit the institutional context:

The mission of North Carolina State University, as stated in the undergraduate catalog and on the university's Web site is "to serve its students and the people of North Carolina as a doctoral/research-extensive, land-grant university. Through the active integration of teaching, research, extension, and engagement, North Carolina State University creates an innovative learning environment that stresses mastery of fundamentals, intellectual discipline, creativity, problem solving, and responsibility. Enhancing its historic strengths in agriculture, science, and engineering with a commitment to excellence in a comprehensive range of academic disciplines, North Carolina State University provides leadership for intellectual, cultural, social, economic, and technological development within the state, the nation, and the world."

Stressing the three-pronged mission of the University, the mission of the College of Engineering is "to provide students with a sound engineering education, advance the understanding and application of scientific principles, enhance economic development, and improve the quality of life of our citizens through teaching, research, and outreach programs. In addition to ensuring that our students are exposed to modern engineering principles and have access to modern equipment and technology to support their educational experience, the College seeks to create a team-oriented environment throughout our academic enterprise. Our goal is to produce well-rounded engineers who can function effectively in the technical arena as well as possess the skills to assume leadership roles in industry, academia, and government." The vision is "to become a nationally renowned college of engineering that is recognized for its outstanding education, research, and outreach programs, and for the quality of its graduates."

The Department of Civil, Construction and Environmental Engineering administers three undergraduate degree programs: Civil Engineering, Construction Engineering and Management, and Environmental Engineering. The mission of the department, which embraces the threefold mission of both the college and university, is "to provide undergraduate and graduate educational programs which prepare graduates to enter into successful professional careers, to increase knowledge and its application through research, and to offer service and life-long learning opportunities through outreach programs."

APPENDIX B:
First Draft of Rubric

Lab Report shows that the student . . .

- can design civil engineering laboratory and field tests in terms of collecting and measuring samples, where to take measurements and how many measurements to take
- possesses a solid understanding of accuracy and precision
- can conduct laboratory or field tests properly and
- can describe the procedures used to conduct those tests so that others can understand them
- can measure and record raw test data and analyze those data for the purposes of understanding and explaining the data
- can represent data in both verbal and visual forms (equations, tables, graphs, figures, etc.) in a way that is both an accurate and an honest reflection of the data
- can make sound judgments in the solution of problems in civil engineering and support those judgments and solutions according to standards of practice in the field

(Appendix C follows on next page)

APPENDIX C:
Final Draft of Rubric

Date:

Course:

Lab:

Students should demonstrate their ability to:

design civil engineering laboratory and field tests:	N/A	N/I	good	excellent
state the objectives of the lab or field test clearly and concisely				
describe the engineering context of the lab				
establish appropriate sampling procedures, e.g., where and how many samples to take				
demonstrate a solid understanding of issues of accuracy and/or precision and their specific impact on the results				
conduct civil engineering laboratory or field tests properly and safely:				
clearly describe the testing procedures followed so that professionals in the field can fully understand them				
show that they are fully aware of and can adhere to the measurement techniques appropriate to meeting the objectives				
analyze and interpret data from civil engineering laboratory and field tests:				
gather appropriate and sufficient data to meet the objectives				
apply appropriate analytical tools to the data to meet the objectives				
show that they can capably represent the data in visual forms (equations, tables, graphs, photographs, etc.)				
show that they can capably represent the data in verbal form by referring to and adequately addressing the visual representations				
demonstrate that they can use the data to make sound interpretation and reasonable conclusions				

ACKNOWLEDGMENTS

I would like to thank James M. Nau and Mohammed A. Gabr of the Department of Civil Engineering at North Carolina State, chairmen of the ABET and Laboratory Committees, respectively, for their valuable cooperation in this project. I am also grateful to my colleagues in North Carolina State's Campus Writing and Speaking Program, Chris Anson and Deanna Dannels.

REFERENCES

ABET: Engineering Accreditation Commission. (2004a, May 13). *Self-Study Questionnaire.* Retrieved June 3, 2004, from
http://www.engr.ncsu.edu/abet/criterion-2/guidelines-2.html

ABET: Engineering Accreditation Commission. (2004b, May 13). *Guidelines to institutions, team chairs and program evaluators on interpreting and meeting the standards set forth in criterion 3 of the engineering accreditation criteria.* Retrieved June 3, 2004, from
http://www.engr.ncsu.edu/abet/criterion-2/guidelines-2.html

ABET: Engineering Accreditation Commission. (2004c, May 13). Criteria for Accrediting Engineering Programs. Retrieved June 3, 2004, from
http://www.engr.ncsu.edu/criterion-2/guidelines-2.html

Carter, M. (2002). A process for establishing outcomes-based assessment plans for writing and speaking in the disciplines. *Language and Learning Across the Disciplines, 6,* 4–29.

Carter, M. (2001a). *ABET criterion 2.* North Carolina State College of Engineering. Retrieved June 28, 2004, from
http://www.engr.ncsu.edu/abet/criterion-2/criterion-2.html

Carter, M. (2001b). *ABET criterion 3.* North Carolina State College of Engineering. Retrieved June 28, 2004, from
http://www.engr.ncsu.edu/abet/criterion-3/criterion-3.html

Commission on Colleges, Southern Association of Colleges and Universities. (2004, February 23). *Principles of accreditation: Foundations for quality enhancement.* Retrieved June 28, 2004, from http://www.sacscoc.org/principles.asp

Hovde, M. R. (1999, October). Enhancing the future of technical communication in schools of engineering and technology: An argument for involving e&t faculty in communication assessment. *Proceedings of the Council on Programs in Technical and Scientific Communication,* 71–74.

Hovde, M. R. (2003a, April). How workplace and academic engineering written communication abilities differ: A foundation for communication assessment. *Proceedings of the Illinois Indiana Sectional Conference of the American Society of Engineering Education,* Valparaiso, IN.

Hovde, M. R. (2003b, June). Assessing engineering and technology students' abilities to 'communicate effectively': Overcoming obstacles. *Proceedings of the National Conference of the American Society for Engineering Education.* Nashville, TN.

Williams, J. M. (2001) Transformations in technical communication pedagogy: Engineering, writing, and the ABET engineering criteria 2000. *Technical Communication Quarterly, 10*, 153–173.

CHAPTER 8

Beyond Denial: Assessment and Expanded Communication Instruction in Engineering and Professional Programs

Steven Youra
California Institute of Technology

In "Expanding the Role of Technical Communication through Assessment," Michael Carter deftly shows how technical communication instructors can play a significant role in the engineering accreditation process, particularly at the program and college levels (Carter, 2009). He illustrates how they can collaborate with engineering faculty on both substance (articulating mission statements, evaluating the success of learning objectives) and methodology (developing and guiding effective and efficient procedures). This object lesson from engineering has implications for outcomes-based assessment across other academic disciplines, and from the program level to the locus of classroom praxis, especially across a curriculum that may incorporate writing, speaking, and visual communication, to achieve specific teaching/learning goals. Carter's piece invites reflection on relationships among outcomes-based assessment methods, writing/communication instruction, and assignment design across engineering and professional programs, both within and beyond the classroom.

FINALLY, ACCEPTANCE

Before recent changes in engineering assessment, faculty and administrators had long criticized what many perceived as an inflexible system imposed by the Accreditation Board for Engineering and Technology, ABET. Under the former assessment process (pre-2000), engineering programs were required to produce meticulous tallies and elaborate justifications about which particular courses

113

contained how many units of various designated components (e.g., engineering design, humanities, engineering science, etc.). In preparing for program reviews, administrators were sometimes driven to creative accounting practices (allocating a quarter-unit of design here, a half-unit there) that rivaled those of a Wall Street investment bank. "The rigidly prescriptive nature of the system was a source of frustration to engineering faculty and administrators, and the derisive label 'bean counting' almost invariably arose whenever ABET came up in conversation" (Felder & Brent, 2003, p. 9).

By contrast, the revised assessment scheme, Engineering Criteria (EC) 2000, was designed to empower each program to define its own mission and to stimulate curricular freedom and experimentation with ways to achieve that mission. The result of a collaboration between academia and industry, this "outcomes-based" procedure, adapts terms and processes from engineering and manufacturing and applies them to education; it uses inputs, outputs, and feedback loops to produce continuous educational improvement though a process of Total Quality Management. If meant to be liberating, EC 2000 was also intended to include a degree of "reflective practice" (Schön, 1983) about the goals and results of teaching and learning—reflection that has proved to be challenging for many programs.

In retrospect, the new assessment method can be seen as a classic case of the situation "Be careful what you wish for, because you just might get it." As Carter (2009) notes, the freedom/obligation for professors and administrators to formulate their own educational objectives and to measure the outcomes has "elicited the most consternation among engineering faculty preparing for ABET reviews" (p. 90). Indeed, the transition to the new assessment process caused strong consternation at many schools. Several years ago, when I worked with an engineering curriculum committee at another university, our committee chair attended a national workshop in preparation for an ABET assessment. On returning, he reported with grim humor that colleagues across the country were all experiencing the same sequence of responses to EC 2000: Denial, Anger, Bargaining, Depression, and finally, Acceptance.

WRITING OUTCOMES IN THE PROFESSIONS

The drive toward more effective assessment of undergraduate education is strongest in engineering and other programs that grant degrees in fields of professional practice. It's not hard to understand why this is the case, since graduates of such programs have a critical impact on the general public—on our health, welfare, and economy. Architects or speech therapists (for example) are liable for the consequences of their work in ways that historians or philosophers are not. (Society expects buildings to have stable foundations, designed by graduates of accredited civil engineering programs; people are less disturbed by, say, destabilizing interpretations from literary critics.) Certification of a professional program is meant to ensure that its graduates have learned enough material and gained sufficient expertise in the discipline to move to the next level of apprenticeship or practice.

Engineering is but one of many fields that have recently shifted assessment procedures from "bean counting" the number of courses and instructional modules on narrowly prescribed topics, to a process of measuring learning outcomes that conform to the mission and goals (the "objectives," in ABET language) of a particular program and its institution. Consider, for example, the revised accreditation standards for programs in nursing that took effect on January 1, 2005:

STANDARD III. PROGRAM QUALITY: CURRICULUM AND TEACHING-LEARNING PRACTICES

The curriculum is developed in accordance with the mission, goals, and expected outcomes of the program and reflects professional nursing standards and guidelines and the needs and expectations of the community of interest. There is congruence between teaching-learning experiences and expected outcomes. The environment for teaching, learning and evaluation of student performance fosters achievement of the expected outcomes.

Key Elements:
III-A. The curriculum is developed, implemented, and revised to reflect clear statements of expected student learning outcomes that are consistent with professional nursing standards and guidelines and congruent with the program's mission, goals, and expected outcomes.
(American Association of Colleges of Nursing, 2003a)

In a related handbook that compares these revised standards with the former assessment criteria, the Commission on Collegiate Nursing Education (CCNE) explicitly highlights the new emphasis on outcomes:

Throughout the [revised] document, all references to "mission, goals, and objectives" and "mission, philosophy, and goals/objectives" were replaced by the phrase **"mission, goals, and expected outcomes."**
(American Association of Colleges of Nursing, 2003b;
emphasis in original)

Similarly, in the field of social work, recently revised assessment standards require educational programs to develop clear and measurable outcomes that are routinely analyzed to enhance the curriculum:

8. Program Assessment and Continuous Improvement

8.0 The program has an assessment plan and procedures for evaluating the outcome of each program objective. The plan specifies the measurement procedures and methods used to evaluate the outcome of each program objective.

8.1 The program implements its plan to evaluate the outcome of each program objective and shows evidence that the analysis is used continuously to affirm and improve the educational program.

(Council on Social Work Education, 2002, p. 17)

These criteria, which took effect in 2003, replaced standards from 1994—older standards in which the word "outcome" never appeared.

As undergraduate programs in the professions are increasingly required to formulate and judge writing and speaking outcomes, technical communication instructors can play an important part in the assessment of those outcomes. These communication specialists may provide expertise with the broader assessment process, but, as Carter (2009) points out, "It may be more feasible for technical communication teachers to focus only, or at least initially, on Criterion 3-g, 'an ability to communicate effectively.' Engineering faculty need help in defining in operational terms what it means for students to write effectively and how to measure that ability" (p. 90). Although faculty committees may focus on program-level rubrics and outcomes related to communication, they must attend to the local contexts within which the rubrics operate—the day-to-day course agendas that include some mix of reading, writing, discussion, computation, experimentation, and other learning activities.

At the course level, a rubric may be fine-tuned to match the precise goals of a particular writing assignment, but if the tangible evidence of an outcome (students' technical reports, for example) veers wildly away from the ideal described by a rubric's analytic scoring scale, then teachers should not only revisit the rubric, but also reflect on their instructional methods, and consider how to produce more effective teaching and learning (improved outcomes). As engineering educators Felder and Brent (2003) explain, the ABET process should actually motivate two kinds of response: "If the assessment reveals that an objective has not been satisfactorily achieved, the nature of the failure may suggest reframing the objective or modifying the instruction used to address it" (p. 9). In other words, if most students' writing in an engineering course does not measure up to expectations, then instead of tweaking the rubric again, it may be time to reconsider how writing is taught within that course. Technical communication instructors are well positioned to help "modify instruction" and improve pedagogy—not only in their own courses (courses that typically enroll large numbers of engineering students), but also within science and engineering courses that include a communication component. Of course, the mere presence of writing or speaking assignments in an engineering or other technical/scientific course does not, in itself, constitute "teaching" communication. To promote effective instruction, assignments must be well designed from a rhetorical as well as a technical standpoint; they must also be integral to the course material, sequenced logically, and accompanied by appropriate guidance and instructional materials.

WAC, WID, CAC, ECAC, Etc.

Too often (and with the best of intentions) the standard formula for including tech comm. in engineering and professional curricula was simply "add writing and stir." Historically, courses with writing have included overly generic assignments with scant guidance ("A research paper is due at the end of the term"). Often, the writing component has suffered from insufficient planning about both its purpose (beyond the school task of demonstrating knowledge or work accomplished) and process (the interim steps that would bring students from rough ideas to a final written product). For example, an assigned lab report may have required students to produce a Conclusions section, when, in fact, the experiment itself (e.g., how to use a new piece of equipment) may not have lent itself to drawing authentic conclusions. In such circumstances, students have been compelled to mold their writing to fit an inflexible template. Unfortunately, the situation ensured that "report writing [is] primarily an exercise, isolated from any sort of meaningful problem-solving context" (Kalmbach, 1986, p. 179). However, the new ABET approach, along with the development of Writing Across the Curriculum (WAC) programs, has stimulated reevaluation of the relationship between writing and learning.

Over the past 30 years, many campuses have developed initiatives that formally integrate communication into subject-matter courses in academic fields, under the auspices of WAC or its acronymic variations, WID (Writing in the Disciplines), CAC (Communication Across the Curriculum), ECAC (Electronic Communication Across the Curriculum), and others. For our purposes here, differences among these endeavors are less important than their shared assumption that effective incorporation of writing and other modes of communication can enhance students' understanding of the subject matter and improve their ability to communicate that understanding.

Currently, in about one-third of top-ranked engineering schools, the technical faculty collaborate with writing instructors to integrate communication into the engineering curriculum (Reave, 2004). When communication teachers work with assessment in engineering and professional fields, the effort does not typically begin at the program or college level, but in the classroom, where the writing meets the road. Such a teaching partnership might start with an invitation (Sageev, Bernard, Prieta, & Tomanowski, 2004, p. 14): An engineering instructor asks a communication teacher to "give a lecture on writing to my thermodynamics class." Alternatively, instructional collaborations might follow from a faculty-development seminar held by a Writing Across the Curriculum (WAC) program, and lead, for example, to ongoing instructional support in a particular science class or to a TA training plan for communications assignments in a technical course (perhaps to bolster a designated "writing-intensive" engineering class). In designing the invited "lecture," or in working with TAs, a technical communication specialist engages outcomes-based assessment issues on the ground level of individual assignments, teaching methods, and course goals. For example, the communication instructor might transform an invited performance into an interactive workshop, in which students might discuss examples, produce drafts, review the work of peers, discuss sample reports, and so on. In a similar vein, a TA training session might include a critique of the lab report genre and discussion of alternative rhetorical tasks that would achieve specific teaching/learning goals.

Such efforts to integrate writing into subject-matter courses may complement existing stand-alone technical communication courses—courses based in a range of possible locations, depending on the institution (e.g., in the engineering college itself, in the English department, or as part of a universitywide writing program). As WAC consultant, a technical communication specialist can provide expertise in

- designing and sequencing assignments that have a communication component;
- developing appropriate instructional materials for such assignments;
- devising methods for giving students effective feedback on writing and speaking effectively and efficiently without creating an unmanageable paper burden;
- planning and (co-)teaching writing workshops associated with technical courses to prepare students for writing and/or speaking assignments in the technical class;
- formulating effective strategies for assessing the communications component (back to outcomes!).

Writing goals and learning goals comingle in the process of careful assignment design and related instruction. As writing and engineering instructors construct assignments together, they inevitably examine what, specifically, they want students to learn and how. According to Sageev et al. (2004), "When technical communication professionals work one-on-one with engineering experts to define, articulate, and document critical-to-quality communication issues peculiar to each discipline and communication task, the genuine process of continuously improving 'outcomes' begins in earnest" (p. 144). The Appendix to this chapter offers guidelines for developing assignments that interweave rhetorical and conceptual aspects. An outcomes-based assessment plan should embody such linkages.

ASSESSMENT AND AUDIENCE

One critical component of a "meaningful problem-solving context" is *audience*, and any assessment of communication in professional programs should include it. After all, a document or presentation is successful insofar as it connects with its intended readers or listeners; its effectiveness can't be evaluated in isolation from the social context. Yet, while some engineering assignments may clearly designate an audience and purpose—for example, design reports aimed at actual clients or industrial partners outside of the classroom, or research papers targeted for professional journals—most conventional engineering assignments do not mention audience at all, and neither does the ABET "effective communication" Criterion 3-g.

Technical communication specialists, ever concerned with audience, should bring this consideration to their work with engineering and professional fields, whether the writing/presentation tasks are situated within or beyond the classroom. Assignments that require writers to address appropriate hypothetical readers (e.g., a nontechnical client who must base a decision on the student's analysis or a manager with an MBA who must be persuaded to fund a proposal, etc.) can sensitize students to important audience factors. In responding to the writing produced for such assignments, instructors should take the audience's point of view, and any rubrics or evaluation criteria should include audience considerations. But even when the assignment specifies a context, student writers may be confused by the disjunction between a fictitious workplace reader and the classroom judge who actually reviews (and grades) the writing. As Dorothy Winsor observes, "Classroom instruction alone can never completely prepare a student to write at work" (1996, p. 20).

Technical communication instructors can also help professional programs move beyond hypothetical audiences to engage with others outside of the classroom through activities such as job internships, co-op placements, client-based projects and service-learning activities. Such experiences give students an opportunity both to "transfer" classroom lessons and to expand their learning in "meaningful problem-solving context[s]" (Kalmbach, 1986, p. 179). For example, in noting that a classroom and a work site represent qualitatively different contexts for learning, some researchers and educational theorists have examined how novices develop authentic professional skills by participating within "communities of practice" (Lave & Wenger, 1991; see also Freedman & Adam, 1996; Katz, 1998; Wenger, 1998). Technical communication instructors can facilitate initiatives that help students learn how to present information effectively within their future professional community, where the audience comprises technical peers, colleagues trained in other fields, and managers with varying degrees of technical understanding.

Through internships in engineering and other professions, students learn about complex audiences before they leave school. For example, students at Cornell

University may fulfill an engineering college writing requirement through work done at a co-op job placement. Following explicit guidelines, the students propose a plan that describes the documents to be produced, feedback from supervisors, and opportunities for revision; they submit several progress reports during the placement period and a final report and portfolio of documents at the end. All elements of the plan, including assessment, are carefully coordinated between workplace mentors and the college's communications program (Cornell University, 2009).

A different shift in audience and writing assessment has students assuming a professional role in work with nonspecialists. For example, at Northwestern University, all first-year engineering students take a two-quarter sequence in Engineering Design and Communication (EDC). Cotaught by instructors from engineering and the campus writing program, these students work in teams to solve real problems for real clients (Northwestern University, 2009). At the University of Southern California, students collaborate on consulting projects primarily for local nonprofit organizations as part of a required course taught by the Engineering Writing Program. In producing documents that benefit these organizations (e.g., recommendation reports, feasibility studies, grant proposals), the students learn to adapt their communications to authentic, responsive, audiences with whom they actually interact (University of Southern California, 2009). Some of the USC projects also involve cross-cultural communication with clients in other countries. An intercultural dimension is also a factor in a project course at the California Institute of Technology: In Design for the Developing World, students work in multidisciplinary, international teams to create inexpensive, sustainable solutions to problems in Guatemala and other countries (recent projects include a basic water purification system for rural villages, a pedal-powered cell phone charger, a simple and efficient tool for separating corn kernels from cobs). The diverse teams include not only engineering students from Caltech but also undergraduates from a nearby Pasadena design college and engineering students from a Guatemalan university. To plan and coordinate their efforts, the teams use a range of communications tools, including e-mail, video conferencing, Skype, and Wikis (California Institute of Technology, 2009).

As audiences for writing assignments shift from instructors and classroom peers to co-workers at different technical levels and to clients near and far, students encounter new and authentic circumstances for demonstrating ABET's EC2000 Criterion 3-g, "an ability to communicate effectively." They must learn to refashion generic conventions to accommodate actual contexts and readers, and to produce documents and presentations that have concrete results. Compared with classroom assignments, communicating in organizations involves more improvisation, provides less consistent feedback, and includes tasks that are not formulated and sequenced in relation to the writer's growing capabilities (Freedman & Adam, 1996, pp. 412–418). As David Russell reminds us,

Learning to write and writing to learn are valuable in so far as they help us and our students to do important things with others, not only in school but beyond school, to make a difference in the worlds students will enter—and eventually remake. . . . Active learning means expanding our students' and our own involvement with other people, with powerful social practices— disciplines, professions, institutions, communities, organizations of all kinds where writing can be transformed and transformative. (1997, pp. 4, 5)

Assessments must take into account these differences in sites of learning and attend closely to audience factors. In the end, both teaching and assessment practices must respond to the many ways in which effective communication *matters*.

APPENDIX
Considerations for Designing Writing Assignments:
Outcomes-Based Assessment Strategies Begin Here

When you design formal writing assignments, consider . . .

Your teaching goals, immediate and long range
- How will the writing serve your teaching aims?
- When students finish this assignment, what do you want them to have learned?
- How might two or more assignments be sequenced toward greater conceptual or rhetorical complexity? Should you divide a large, ambitious writing/thinking task into several smaller portions that lead to the final product?

Explicit or implicit assumptions
- In what ways is the assignment intended to bring students from the known to the unknown?
- What are the assignment's implicit assumptions about what students need to know in order to perform adequately? Do they know what they need to know?
- Who is the intended audience for the assignment?
- In your view, does the assignment have clearly right/wrong answers or does it have a range of analyses, interpretations, or conclusions? Does the wording of the assignment indicate which of these is the case?

Relationships between conceptual task(s) and wording of the assignment
- What is required by the various writing/thinking tasks; e.g., describe, discuss, analyze, compare/contrast, explain, justify, examine, assess, interpret persuade, summarize, determine, etc.?
- How many procedures or steps are built into this assignment? Are students equipped to do them? From the wording of the assignment, will they understand that they are to do them all?

- What is the relationship between these steps and the writing format?
- Is there an organizing principle or prescribed format that students should use?
- Will students know what sorts of data, visuals, and support to use?

How you will prepare students to do the assignment well
- Would examples (good or bad) clarify what you expect of the students?
- Or is the assignment meant to challenge students to formulate their answers with little guidance?
- What do you need to explain about the relationship between the technical content and the writing? To what extent are these elements separable?
- Will students do the writing in stages, handing in one or more early drafts or other preliminary writing (e.g., lists of ideas, brainstorming material, outlines, proposals)? If so, what forms of interim feedback will the writing receive from you or their peers?
- Do students understand the criteria by which their performance will be evaluated? What is the mechanism for feedback and revision?
- Will students be able to revise their paper? How will revisions be factored into the grade?

When you present formal writing assignments . . .
- Prepare a **handout** that describes the task and announces the due date, format, and suggested length.
- Include information that students will need in order to meet **your expectations**. Anticipate problems, offer tips, and make explicit your answers to questions listed above.
- Consider producing a **checklist** of key factors that students can use when writing, and that you can use when grading.
- Set aside class time to **discuss** the assignment and its relationship to the course. Do not assume that students understand its purpose or direction. Consider creating (improvising with students in class) a sample outline for or example of a response to the assignment.

REFERENCES

American Association of Colleges of Nursing. (2003a, October). *Standards for accreditation of baccalaureate and graduate nursing programs.* Retrieved May 1, 2009, from http://www.aacn.nche.edu/Accreditation/NEW_STANDARDS.htm

American Association of Colleges of Nursing. (2003b, October). *Understanding the changes to the CCNE standards for accreditation.* Retrieved May 1, 2009, from http://www.aacn.nche.edu/Accreditation/PDF/UnderstandingChanges.pdf

California Institute of Technology. (2009). *E105: Product design for the developing world.* Retrieved May 1, 2009, from http://www.its.caltech.edu/~e105/

Carter, M. (2009). Expanding the role of technical communication through assessment: A case presentation of ABET assessment. In J. Allen & M. Hundleby (Eds.),

Assessment in technical and professional communication (pp. 89-111). Amityville, NY: Baywood.

Cornell University College of Engineering. (2009). *Writing-intensive co-op.* Retrieved May 1, 2009, from http://www.engineering.cornell.edu/programs/undergraduate-education/engineering-communications/tech-writing-program/writing-intensive-co-op.cfm

Council on Social Work Education. (2002). *Educational policy and accreditation standards.* Retrieved June 1, 2006, from http://www.cswe.org/accreditation/EPAS/EPAS_start.htm#f2

Felder, R. M., & Brent, R. (2003). Designing and teaching courses to satisfy the ABET engineering criteria. *Journal of Engineering Education, 92*(1), 7–25.

Freedman, A., & Adam, C. (1996). Learning to write professionally: 'Situated learning' and the transition from university to professional discourse. *Journal of Business and Technical Communication, 10*(4), 395–427.

Kalmbach, J. R. (1986). The laboratory reports of engineering students. In A. Young & T. Fulweiler (Eds.), *Writing across the disciplines: Research into practice* (pp. 176–183). Upper Montclair, NJ: Boynton/Cook.

Katz, S, (1998). Part II—How newcomers learn to write: Resources for guiding newcomers. *IEEE Transactions on Professional Communication, 41*(3), 165–174.

Lave, J., & Wenger, E. (1991). *Situated learning: Legitimate peripheral participation.* Cambridge, MA: Cambridge University Press.

Northwestern University School of Engineering and Applied Science. (2009). *Engineering design and communication.* Retrieved May 1, 2009, from http://www.segal.northwestern.edu/undergraduate/edc/

Reave, L. (2004). Technical communication instruction in engineering schools: A survey of top-ranked U.S. and Canadian programs. *Journal of Business and Technical Communication, 18*(4), 452–490.

Russell, D. (1997). Writing to learn to do: WAC, WAW—Wow! *Language and learning across the disciplines, 2*(2), 3–8. Retrieved May 1, 2009, from http://wac.colostate.edu/llad/v2n2/russell.pdf

Sageev, P., Bernard, K., Prieto, F., & Romanowski, C. (2004). *Safe passage through the engineering curriculum: Guiding subject experts toward integration of communication instruction and outcomes assessment.* Proceedings of IPCC 2004. IEEE Professional Communication Conference, 139–146.

Schön, D. A. (1983). *The reflective practitioner.* New York: Basic Books.

University of Southern California Viterbi School of Engineering. (2009). *Community consulting projects.* Retrieved May 1, 2009, from http://viterbi.usc.edu/academics/programs/ewp/community/

Wenger, E. (1998). Communities of practice: Learning as a social system. *Systems Thinker 9.5* Retrieved May 1, 2009, from http://www.co-i-l.com/coil/knowledge-garden/cop/lss.shtml

Winsor, D. (1996). *Writing like an engineer: A rhetorical education.* Hillsdale, NJ: Erlbaum.

Assessing the Work of Graduate Students

CHAPTER 9

Assessment of Graduate Programs in Technical Communication: A Relational Model*

Nancy W. Coppola and Norbert Elliot
New Jersey Institute of Technology

In this chapter, we offer a consolidated model for the assessment of graduate programs in technical communication. Because we are empiricists—believers that inductive methods, in which detailed and multiple observations are made, prove more effective in organizational settings than individual statements of faith—our chapter takes the standard format of a study. We begin with background of our graduate program in professional and technical communication at the New Jersey Institute of Technology (NJIT, 2007) and the relational theory that informs our methods, and we then describe our assessment design, the procedures used to conduct the study, the results of our efforts, and conclusions drawn.

BACKGROUND:
A TRADITIONAL AUDITING MODEL

In spring 2001, the Master of Science in Professional and Technical Communication (MSPTC, 2007) at NJIT initiated an internal review process, as scheduled

*This study was supported in part by a grant from the Council for Programs in Technical and Scientific Communication (CPTSC, 2007).

by the provost and the Review Committee for Department and Program Assessment. We followed the process guidelines, similar to those we had often used to help our university prepare for review by the Accreditation Board for Engineering and Technology (ABET, 2007) and the Middle States Commission on Higher Education (MSCHE, 2007). The evaluation procedure was conducted and reported in routine fashion: We presented the curriculum vitae of faculty, the number of courses taught and grades earned, the student demographics, and our plans for future growth. In that ours is unique among NJIT's graduate programs—students can earn the MSPTC, the University's sole liberal arts graduate program, entirely in a distance-learning environment—the program review drew the expected attention: the supportive engagement of the science and liberal arts dean who found the program innovative, and the sharp criticism from the computer science dean who judged the program feeble. The process was, essentially, a routine audit: the product was a lengthy document written by committee; the process was a staged review performed in a conference room; and the outcome was a final memo certifying that we could continue our recruitment efforts.

This audit, nevertheless, gave us a valuable picture of our MSPTC program and the 12-credit Certificate in the Practice of Technical Communication (2007) serving as a program feeder. Demographically, we have continued to meet the NJIT mission of diversity: 70% of our students are women. Seven full-time, tenured and tenure-track research-active faculty devote 50% of their instructional time to the 71 enrollments in Fall 2007 courses. In benchmarking our program to other leading national programs, ours remains one of the very few Master of Science programs in professional and technical communication in which the entire program can be taken in a distance-learning, asynchronous format. And the MSPTC student evaluations remain among the highest in our Department of Humanities and the University.

As a result of our internal review process, we were able to identify four elements of program review, shown in Figure 1, which may be understood as traditional components used to assess graduate programs in technical communication:

- a gauge of institutional commitment in the faculty lines allocated to the program;
- a more precise understanding of the curriculum and instruction as we benchmarked our program against others;
- a measure of student satisfaction (both rising and falling scores) in the information taken from our course surveys; and
- an understanding that faculty willingly gave their instructional and advising time to the program and, by all accounts, were pleased to work with our students.

These components, understood as static, are assumed to impact the MSPTC program goal: the ability to educate students, thus preparing them to be working

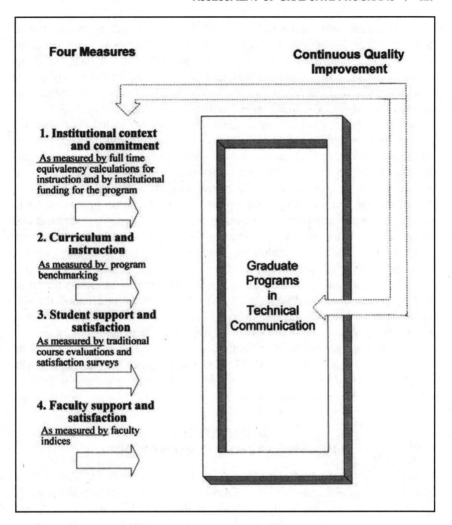

Four Measures

Continuous Quality Improvement

1. Institutional context and commitment
As measured by full time equivalency calculations for instruction and by institutional funding for the program

2. Curriculum and instruction
As measured by program benchmarking

3. Student support and satisfaction
As measured by traditional course evaluations and satisfaction surveys

4. Faculty support and satisfaction
As measured by faculty indices

Graduate Programs in Technical Communication

Figure 1. Traditional assessment model for graduate programs in technical communication.

professionals. This goal, we claim, is to be achieved through the following strategies: requiring work in emerging communication technologies; fostering a critical view that blends theory and practice; providing orientation to our complex, global society in which communication is situated; encouraging wise use of courses in other NJIT graduate programs in engineering, architecture, management, and applied science; and encouraging relationships between industry and the program in order to promote the continuing development of needed professionals in the field.

Yet the traditional model had its limits. While this auditing model led us to identify four components of graduate education in technical communication, the really important questions remained unanswered: Did we really meet the program goals in terms of student performance? How could a valid performance assessment allow us to better understand the use of outcomes assessment in program evaluation? Too many unanswered questions remained.

In 1989, Samuel Messick argued that validation should involve integrated evaluative judgments in support of the "adequacy and appropriateness of infer-ences and actions" based on assessment mode (p. 13). Within an integrated framework, he argued, validity and values must be seen as "one imperative, not two," and thus the consequences of testing must be considered (p. 92). In that same year, Egon G. Guba and Yvonna S. Lincoln called on researchers to extend the empirical traditions of measurement, description, and judgment and to embrace the fourth generation of evaluation in which stakeholder concerns are recognized as significant. Within the new mode of "responsive constructivist evaluation," stakeholders were to be taken as an essential element of the inves-tigative process and the impact of the evaluation was to be best understood within specific contextual settings (p. 38). The better we understood the sig-nificance of "local meaning and utility," Guba and Lincoln believed, the more we might be able to empower those we were striving to serve (1989, p. 47). Within such a framework of validation, we decided that it was time to do better—for the administrators who funded the program, for the faculty who allocated time to teach in it, and for the graduate students who came to us to be educated in technical communication.

ORIENTATION:
A RELATIONAL ASSESSMENT MODEL

In place of a traditional, static model of stimulus and response actions asso-ciated with program assessment (put-resources-in-and-watch-results-appear), we framed the auditing model as both relational (look to association!) and behavioral (look to performance!) in orientation. We knew to value the potential for rich associations, the insistence on detail, and the desire to create a non-mechanistic cycle of improvement.

Relational models have a richly complex history in assessment. In order to infer the presence of intelligence from the relationships of variables such as the ability to perceive subtleties in sensory stimulation and the ability to achieve high marks in French, Charles Spearman (1904a, 1904b) set forth the idea of strength of association between variables. He literally invented the idea of correlation and probability to reify the presence of native intelligence, an invalid construct that continues to haunt modern researchers. Today, philosophers of the social sciences such as Martin Hollis (1994) remind us that a cause is simply an instance of regularity. As H. Russell Bernard (2000) cautions social

researchers, conditions such as covariation, lack of spuriousness, temporal prece-
dence, and robust theoretical underpinnings must each be satisfied to establish
a cause-and-effect pattern (2000, pp. 53–57). Consequently, it is best to speak of
describing relational models—not declaring causal evidence—in studies such
as ours.

In the present study, we therefore note that the model we offer is designed
to be relational due to additional presence for outcomes assessment. Specifically, we
wanted to know how variables of technical communication (X) shown in Figure 2
were related to the performance of students as captured in their e-portfolios (Y).

The outcomes assessment plan was designed, the key questions posed. While
we knew a great deal about results gained by means of the auditing model, our
profession has yet to produce a body of research regarding an outcomes assessment program centered on the e-portfolios of graduate students. Our intention
was to use the evaluated performance of our students to inform and refresh the
aims of our degree program. In place, such a relational framework would allow

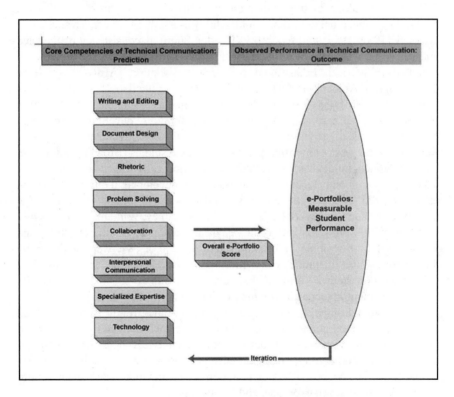

Figure 2. Outcomes assessment plan for graduate student
e-portfolios in technical communication

our students to learn more about their performances from the collective voice of the faculty, a voice that could be validated with empirical information gained through valid and reliable measurement concepts. That is where we dug in.

PROCEDURE IN THREE PHASES

We began by deliberately problematizing the fifth measure, shown in Figure 2, in order to explore its potential to capture the performance of our students. Once that phase was completed, we could then attempt to use aspects of performance as they allowed us to undertake those two archetypal educational activities: formative and summative assessment.

Phase 1: Problematizing Outcomes Assessment

Within the realm of outcomes live all the complexities associated with contemporary writing assessment. Although much has been accomplished in contemporary writing assessment (Huot, 2002; Graham & Perin, 2007), much remains to be done in the evaluation of Web-sensible e-portfolios (Yancey, 2004). Our conceptual definitions hold validity as a unitary concept (Kane, 2006; Messick, 1989, 1994) and reliability (Haertel, 2006) as consistency of repeated measurement. Applications of these definitions were informed by the literature of standards (American Educational Research Association [AERA], American Psychological Association [APA], and National Council on Measurement in Education [NCME], 1999) and that of consequence (Wiggins, 1993, 1994). Although we had a solid set of conceptual definitions, developing the list of variables for the first phase was notably difficult. Although there are lists of variables of professional writing generated for case studies (Broadhead & Freed, 1986), our discipline has no collection of empirically based and nationally recognized core competencies comparable to that developed in other fields (Day, 2003). We did, of course, have access to a deep sense of practitioner knowledge (Harner, Johnson, Rainey, & Rude, 2003; Davis, Ramey, Williams, Gurak, Krull, & Steehouder, 2003), as well as survey data (Dayton & Bernhardt, 2004) and categorized bibliographic information (Alred, 2003). We were additionally informed by the literature of writing program evaluation from the field of composition studies (Witte & Faigley, 1983) and by calls for contextually based assessment in technical communication studies (Allen, 1993).

Our research design for a detailed examination of the fifth measure, delineated in Table 1, thus incorporated a set of eight competencies, derived from survey and bibliography, for assessing the outcome of graduate-level technical communication: writing and editing; document design; rhetoric; problem solving, professional traits, and work skills; collaboration and teamwork; interpersonal communication, specialized expertise; and technology.

The relationships between Figures 1 and 2 are significant far beyond their visual value. While the four variables provided in Figure 1 constitute a traditional model for the assessment of graduate *programs* in technical communication, these eight competencies, treated as independent variables in Figure 2, are the basis of our relational model for the assessment of graduate student *ability* within programs. Commitment of institutional resources, curricular and instructional design, student satisfaction and support, and faculty support and satisfaction are thus to be seen in a relational context to student performance—all of which, taken together, enable an assessment of our MSPTC. Thus, student assessment is understood as part of a consolidated, dynamic program assessment model, not as an additional element of testing that may interrupt student progression through the program, a barrier model that we reject. Rather, assessment of the e-portfolios provides a snapshot—one with, as we will demonstrate, very high resolution—of our students as they are understood within a programmatic framework.

To validate this student-centered assessment model within our own NJIT context, we attended to Brian Huot's (2002) call "to emphasize the context of the texts being read, the position of the readers and the local, practical standards that teachers and other stakeholders hold for written communication" (p. 104). Huot's principles for the next generation of writing assessment are site based, locally controlled, context sensitive, rhetorically responsive and shareholder accessible. No matter how far our profession advances in the development of national standards (Dayton, Davis, Harner, Hart, Mueller, & Wagner, 2007), we recognize that those standards will have to be localized if they are to be of any use to those attempting to evaluate particular graduate programs. We thus asked NJIT's MSPTC program instructional faculty to review the criteria that would be used to assess the core competencies and to particularize them, describing how a cognitive skill area like writing and editing, for example, might be addressed in their courses. As well, we wanted to learn if these critical skills were being taught in our program courses. We were especially interested in addressing instruction in the emerging area of visual literacy, comprehensively conceived as a set of techniques that Fleckenstein (2004) terms polymorphic literacy—those strategies of "reading and writing that draw on verbal and nonverbal ways of shaping meaning" (p. 613).

Early in the spring 2004 semester, MSPTC faculty met for practitioner validation. Colleagues arrived with lists of how the core competencies we detailed were adapted within the NJIT curriculum. The faculty sought to define these competencies through questions such as the following:

- What do we really value regarding writing and editing? What are appropriate design elements for Web-based communication?
- How may these elements be determined through usability analysis? What do we talk about when we talk about professional ethics?
- How can group leadership be demonstrated in an online course?

Table 1. Core Competencies for Graduate Students in Technical Communication

Trait	Descriptor	NJIT Descriptor
Writing and editing	Style, correctness, organization	• Demonstrates accurate, thorough, relevant, and coherent content and ideas • Demonstrates accurate language (usable, grammar, punctuation, spelling) • Exhibits clear style (readable, euphonious, concise, cohesive) • Is capable of adapting tone for audience and purpose • Demonstrates cohesion by graphic means (headings, white space) • Demonstrates ability to develop content for Web-based applications • Can promise multiple drafts • Is capable of integrating instructor and peer review comments • Is able to incorporate macro/micro editing
Document design	Visual communication, format, graphics, usability, user-centered design	• Employs technical and aesthetic artistic appreciation • Demonstrates command of print format, graphics, visual communication and assurance of their usability • Demonstrates ability to discern appropriate design elements for Web communication (color, white space, navigation structure, contiguity) • Exhibits ability to incorporate multimedia into online presentations • Demonstrates ability to create user-centered layout and design • Demonstrates understanding of typography nuances
Rhetoric	Audience analysis, ability to adapt communication to context and purpose, genre knowledge, rhetorical problem solving, cultural interpretation, reflective practice, disciplinary history	• Demonstrates understanding of rhetorical strategies and communication theory or learning theory • Capable of conducting user and task analyses in the field and needs analysis for training • Is able to develop and adapt content for audiences • Demonstrates knowledge of persuasive techniques • Exhibits knowledge of disciplinary history

Category	Description	Indicators
Problem solving, personal traits, work skills	Critical thinking, analysis, flexibility, ethics, organization, humor, ability to learn, professionalism, attention to detail, time management, cultural awareness, diversity, lifelong learning	• Demonstrates understanding of diversity and capable of cross cultural and gender communication • Demonstrates competent decision making and problem solving • Capable of understanding complex ideas and written analyses • Shows ability to plan using time line and Gantt chart • Integrates understanding of professional ethics • Capable of analyzing various professional contexts • Understands author-editor relationships • Understands strategies and skills for long-term learning
Collaboration and team work	Group leadership within complex organizational settings	Demonstrates group leadership within complex organizational settings: • Team projects • Online class discussions • Consensus building • Web projects • Conflict resolution
Interpersonal or oral communication	Presentation skills, interviews, listening	• Able to interview subject-matter expert • Demonstrates command of interpersonal communication • Capable of making informal presentations • Understands interview as tool in needs analysis • Develops "alternate" synchronous and asynchronous communication skills using the various WebCT tools like chat, discussion, e-mail
Specialized expertise	Research skills (ability to do research; familiarity with research literature), project management, knowledge management, business practices, scientific and technical knowledge	Demonstrates research skills in the following areas: • Observation methods • Literature review • Bibliographic review • Research funding sources • Content research for manual • Content research for editing • Business practices
Technology	Facility with, critical understanding of, and ability to earn technologies	Exhibits competences in the following areas: • Microsoft Office applications (including Vision, Excel for statistics), Web design tools, and graphics programs • Web site building, XML, HTML • Web-based training and online help authoring tools

- What interpersonal skills are associated with effective asynchronous communication?
- How does the academic world of scholarly research inform the applied world of the workplace? How do we educate—not merely train—students to use technology?

The authors reviewed the taped transcripts of the meeting and met to review the individual, faculty-generated matrices, sorting them according to common categories. We posted these into the core matrix as NJIT descriptors that we then sent back to instructors for verification, elaboration, and identification of missing pieces. At the same time, we met with a member of our corporate advisory board to review the criteria from yet another stakeholder perspective. Over time, we thus derived the NJIT descriptors shown in Table 1 in our quest for complete construct representation.

With a program assessment framework focused on graduate student performance, we were ready to evaluate the ways those competencies were taught within the NJIT curriculum. In our design, we had been responsive to Messick's emphasis on unified validation and Guba and Lincoln's call for attention to stakeholders. We were meeting Wiggins's call for due process and Huot's call for contextualism. We were ready to move on to formative assessment.

Phase 2: Conducting Formative Review

We wrote to our students in the spring of 2004 and asked for volunteers willing to include their e-portfolios in our assessment study. Seventeen students responded, many of whom already maintained Web-sensible sites. Once mid-semester had passed, the faculty met to evaluate qualitatively the e-portfolios. We provided faculty with a form, derived from the descriptors list, which was to serve as a formative "report card" for each student in the MSPTC program. We asked faculty to complete one formative report card for each student in his/her seminar, paying special attention to the specific ways that students handled the core competencies. With Web pages at the ready, faculty met to undertake a collaborative, formative review. Again, the session was taped and the comments reviewed by the authors. At the end of the session, instructor comments surrounding each of the eight competencies were summarized and sent in e-mail to each of the students. Comments on the formative assessment, derived from the tapes, addressed strengths ("He's the best student in my class right now. Very professional, very diligent. One of my assignments, he did twice. Learning Flash wasn't enough for him so he did a digital video of his daughter.") and afforded areas for improvement ("The student does high quality work but does not take part in online discussion. I always have the feeling he was absent; he just wasn't there."). The tone of the assessment was both encouraging and positive.

As of the present writing, the formative assessment has been conducted three times with positive benefit for the students and their instructors. Chief among these benefits is the formation of a community of assessment. Recent trends in writing assessment have begun to focus on the importance of community (Elbow, 2006; Inoue, 2005), and our work serves to validate the experiences of other researchers. The benefits of our assessment include a clearly demonstrable set of expectations for our students that can be fulfilled in a variety of ways. Students address the core competencies in diverse rhetorical forms, from innovative, multimedia presentations employing podcasts to traditional researched essays. As well, since there is no atmosphere of discipline and punishment—both formative and summative assessments are conducted with the realization that the student has no cut score to reach, no subsequent penalty to pay if a standard is not met—continuous quality improvement is the order of the day. Within such an atmosphere, instructors come to realize that they are undertaking assessment in order to articulate values that are important to them in fostering professional behaviors. The individual instructional voice is therefore only part of the process as an equally important collective voice emerges, informed by scholarly nuance (how aesthetic is achieved in a Web site) rather than by rule (how the student has failed to demonstrate the cohesive use of color). We have not sought, and thus have not achieved, a systematic unity—such was never our aim. Rather, as Glenn Tinder (1980) has observed of community, we have achieved a substantial unity. We are leaning how to inquire together *about* technical communication *with* each other (p. 24).

Phase 3: Conducting Summative Review

At the close of the spring 2004 semester, program faculty met again to review e-portfolios, this time within a quantitative framework. While we agree with Wiggins (1993) that all assessment should be thought of as formative (p. 51), we needed to explore a quantitative model that, we believed could complement the qualitative, formative assessment. If successful, we imagined in 2004—and this has indeed become the present case—each fall, we would complete a formative assessment and each spring a summative one, repeating the process each year. The review of the year's end work was, thus, summative in name only; quantitative review would provide information for further improvement of both student ability and the program itself.

Following a calibration session in which the authors presented e-portfolio student work and asked the faculty to discuss levels of work and agree or disagree with the preliminary ratings assigned by the authors, the instructors each independently read an assigned number of e-portfolios with the criteria expressed in Table 1 before them. As they read the e-portfolios, our colleagues were asked, independently, to analyze each of the eight competencies according to a 6-point Likert scale as these competencies were manifested within the students' work. We followed the distinction made by Purvis, Gorman, and Takala (1988) between primary-trait and analytic scoring. That is, primary-trait scoring, in which the scoring scheme is tied to the task, could not be used because of the varied nature of the tasks across seminars. Analytic scoring, taking into account "the specific features or aspects of writing in relation to a general framework which specifies components that it is thought helpful to distinguish," was thus employed, in that it was more suited to our assessment context (Purvis et al., 1988, p. 46). Thus, under the core competency of writing and editing, the following statement appeared on the evaluation sheet: "The contents of the e-Portfolio demonstrate that the student has competent writing and editing skills, as described in the assessment matrix." After reviewing the e-portfolio and considering the statement in relation to Table 1, the instructor circled one of the following values: very strongly agree (score of 6), strongly agree (score of 5), agree (score of 4), disagree (score of 3), strongly disagree (score of 2), and very strongly disagree (score of 1). Similar questions were included on the evaluation sheet for each of the eight competencies. A final statement, the holistically evaluated overall e-portfolio score, considered the submitted materials as demonstrating superior work in the program (6), very good (5), average (4), below average (3), near failure (2), or work at a level of failure (1). Once an analytic evaluation was completed for each of the competencies and a holistic evaluation for the overall score, the sheet was submitted to the authors. Each e-portfolio received two independent readings. Hence, a score for each competency could range from 12 (the highest) to 2 (the lowest).

READER RELIABILITY, CENTRAL TENDENCY, VARIABLE INTERACTIONS, AND CRITERION RELATIONSHIPS

Four key questions had to be addressed in an analysis of the summative assessment, each revolving around significant issues in contemporary writing assessment: the consistency of the reader observations; the performance of the students; the interaction of variables used to capture performance; and relationship of these variables to a criterion measure external to the present assessment process. These questions are discussed in the following paragraphs:

(1) What was the degree of consistency achieved by the faculty in reading the e-portfolios?

As Achilles heels go in the history of writing assessment, reliability is chronically troublesome (Elliot, 2005, pp. 277–292; Haertel, 2006, pp. 101–102). While there is evidence that traditional portfolios have been read reliably, the readers often are asked to submit only a single score for each portfolio (Hamp-Lyons & Condon, 2000, pp. 134–135; Haswell, 2001, pp. 47–48, 74–79; White, 1994, pp. 300–303). We had asked for eight analytic scores and a holistic score for each student's e-portfolio. Only if the reading had yielded measures of interreader reliability could student performance and interaction within the core competencies be reported. In that little has been established about the ability of readers to achieve consensus (interreader agreement) and consistency (interreader reliability) regarding the assessment of e-portfolios of graduate students in technical communication programs, we decided to apply multiple measures of agreement and reliability, as defined by Stemler (2004), to learn as much as we could about reader behavior. In the following analysis, all data are drawn from four assessment episodes: the spring of 2004 ($n = 17$), the spring of 2005 ($n = 22$), the spring of 2006 ($n = 31$), and the spring of 2007 ($n = 27$).

To begin, we wanted to establish, simply, the cases in which readers agreed in their assessment of each of the eight individual competencies and the overall score. We decided to record lack of agreement if the readers awarded scores that were not adjacent regarding a particular trait (e.g., a score of 5 and a score of 3). These discrepant readings were assigned to a third reader for adjudication. The interreader agreement results requiring no adjudication appear in Table 2.

Overall, we were pleased at the levels of interreader agreement during our first assessment in 2004. Nearly 30% of the e-portfolios needed no adjudication whatsoever, an outcome that is impressive when it is considered that 153 observations were made by at least two readers (17 students evaluated on 8 core competencies plus an overall e-portfolio score). Faculty readers were clearly in agreement on the overall score and the rhetoric score (both achieving 94.1% agreement), as well as on the writing and editing, document design, and specialized expertise scores (each achieving 88.2%). Readers found it more difficult to agree on the technology score (76.5%), the problem-solving score, and collaboration score (each achieving 70.6%); and the lack of interreader agreement on the interpersonal communication score (47.1%) was problematic, especially when it was found that four of the readers could not judge the variable at all. In fact, one reader was not able to judge writing and editing, document design, specialized expertise, and the technology variables; and two readers could not judge the collaboration score. In subsequent years—as we became more certain of our evaluative abilities and the students became increasingly aware of the need to identify each variable within their e-portfolios—we adopted the policy of awarding a score for every variable. If the presence of a variable could not be identified, the e-portfolio was awarded a score of 1 for that variable.

The scores from the spring of 2005 reveal that the readers were indeed becoming more familiar with the assessment. All readers reported that they were able to evaluate the competencies in our second assessment, and the most troublesome competencies to assess—interpersonal communication and technology—were read with far greater agreement, yielding an increase in agreement of 44.8% for the former and 18.8% increase for the latter. Similar to the agreement rate in the previous year, just over 30% of the e-portfolios needed no adjudication whatsoever, an outcome that is impressive considering 198 observations were made by at least two readers in 2005. As the number of e-portfolios

Table 2. Interreader Agreement Requiring No Adjudication,
2004 through 2007

CORE COMPETENCIES	YEAR			
	2004 ($n = 17$)	2005 ($n = 22$)	2006 ($n = 31$)	2007 ($n = 27$)
Portfolios needing no adjudication	$n = 5$ 29.4%	$n = 7$ 31.8%	$n = 8$ 25.8%	$n = 10$ 37%
Writing and editing	$n = 15$ 88.2%	$n = 19$ 86.3%	$n = 22$ 71.0%	$n = 22$ 81.5%
Document design	$n = 15$ 88.2%	$n = 18$ 81.8%	$n = 24$ 77.4%	$n = 25$ 92.6%
Rhetoric	$n = 16$ 94.1%	$n = 19$ 86.3%	$n = 23$ 74.2%	$n = 19$ 70.4%
Problem solving	$n = 12$ 70.6%	$n = 19$ 86.3%	$n = 22$ 71.0%	$n = 21$ 77.8%
Collaboration	$n = 12$ 70.6%	$n = 18$ 81.8%	$n = 27$ 81.7%	$n = 27$ 100%
Interpersonal communication	$n = 8$ 47.1%	$n = 15$ 68.2%	$n = 18$ 58.1%	$n = 23$ 85.2%
Specialized expertise	$n = 15$ 88.2%	$n = 19$ 96.3%	$n = 18$ 58.1%	$n = 20$ 74.1%
Technology	$n = 13$ 76.5%	$n = 20$ 90.9%	$n = 24$ 77.4%	$n = 22$ 81.5%
Overall score	$n = 16$ 94.1%	$n = 19$ 86.3%	$n = 26$ 83.9%	$n = 25$ 92.6%

to be read increased in 2006, the interreader agreement rates generally fell, yet rose again in 2007 as the readers became more accustomed to the increased volume. During that reading, 37% of the e-portfolios read needed no adjudication, and the highest rates of agreement ever were recorded for document design (92.6%), collaboration (100%), and interpersonal communication (85.2%), the last rate of agreement proving especially high when we recall that the lowest rate of agreement on any individual variable during the three year period (47.1%) was awarded to that same variable in 2004.

Along with descriptive measures, we wanted to use inferential statistics regarding levels of reliability to determine our understanding of reader consistency. Because interreader reliability has proven such an obstacle to assessment, we decided to overreport our data by using three measures: a weighted Kappa estimate, a Cronbach's Alpha correlation, and Pearson's product moment correlation. The weighted Kappa (K, an agreement measure for categorical data) and Pearson's product moment correlation (r, a measure of correlation between continuously scaled variables) allowed us to test the significance of the agreement by assuming that there was no relationship (the null hypothesis) in reader agreement unless the level of agreement reached the 95% (.05) confidence level. (That is, to be found statistically significant, all probability values had to be equal or be less than a probability value of .05.) Because this was the first occasion for this type of assessment, as a further guard against Type 1 error, we used a 2-tailed test for all correlations. While Cronbach's alpha (α, a measure of internal consistency based on average interscore correlation) allowed a third measure of reliability, that measure did not allow a confidence interval to be established. Because we did not want to mask the results of the analysis, a practice common when only resolved scores are reported, we decided to report both the nonadjudicated scores and the adjudicated scores. The results of our reliability analysis appear in Tables 3 and 4.

In 2004, as Table 3 shows, the nonadjudicated weighted Kappa achieved the 95% confidence level for rhetoric, specialized expertise, and the overall score, although the correlations for these variables could only be described as fair—according to strength of agreement categories established by Landis and Koch (1977, p. 165)—for the specialized expertise variable ($K = .368$) and moderate for the rhetoric variable ($K = .43$) and the overall score ($K = .472$). Under adjudication, each of the correlations met the 95% confidence interval except for problem solving and interpersonal communication, with a slight correlation ($K = .389$) for writing editing and fair correlations for document design ($K = .435$), rhetoric ($K = .43$), and specialized expertise ($K = .448$). Both collaboration ($K = .564$) and the overall e-portfolio score ($K = .588$) achieved moderate correlations. When Pearson's correlation coefficient was computed for the nonadjudicated scores, rhetoric ($r = .494$), specialized expertise ($r = .528$), and the overall score ($r = .58$) reached the 95% confidence interval, traditionally considered the minimum level for establishing interreader reliability when decisions regarding an individual writer are to be made (Cherry & Meyer, 1993, pp. 134–136). Under adjudication, naturally, the Pearson correlation coefficients rose for the variables of writing and editing ($r = .615$), document design ($r = .587$), rhetoric ($r = .494$), problem solving ($r = .523$), collaboration ($r = .765$), specialized expertise ($r = .66$), and the overall e-portfolio score ($r = .73$); each of these correlation coefficients met the 95% confidence interval. Analysis of the

Table 3. Interreader Reliability, 2004 and 2005

YEAR	2004 (n varies from 13 to 17)						2005 (n = 22)					
MEASURE	Non adj. weigh kappa	Adj. weigh kappa	Non adj. Pear r (2-tailed)	Adj. Pear r (2-tailed)	Non adj. Cron α	Adj. Cron. α	Non adj. weigh kappa	Adj. weigh kappa	Non adj. Pear r (2-tailed)	Adj. Pear r (2-tailed)	Non adj. Cron α	Adj. Cron α
CORE COMPETENCIES												
1. Writing and editing	.31	.389*	.49	.615*	.656	.761	.318*	.5**	.459*	.715**	.627	.843
2. Document design	.391	.435*	.43	.587*	.602	.736	.281*	.488**	.523*	.747**	.674	.854
3. Rhetoric	.43*	.43*	.494*	.494*	.656	.656	.323	.441**	.547**	.691**	.692	.812
4. Problem solving	.048	.279	.128	.523*	.221	.68	.238	.416**	.415	.7**	.579	.819
5. Collaboration	.308	.564**	.512	.765**	.631	.864	.34**	.493**	.55**	.756**	.709	.859
6. Interpersonal communication	-.226	.111	-.33	.278	-.923	.419	.241**	.549**	.423**	.779**	.594	.876
7. Specialized expertise	.368*	.448**	.528*	.66**	.691	.795	.435**	.63**	.542**	.858**	.686	.911
8. Technology	-.071	.076*	.012	.335	-.024	.5	.463**	.506**	.563**	.634**	.711	.767
9. Overall score	.472**	.588**	.58*	.727**	.73	.84	.53**	.568**	.686**	.742**	.813	.852

$*p < .05$; $**p < .01$.

Table 4. Interreader Reliability, 2006 and 2007

YEAR	2006 (n = 31)						2007 (n = 27)					
MEASURE	Non adj. weigh kappa	Adj. weigh kappa	Non adj. Pear r (2-tailed)	Adj. Pear r (2-tailed)	Non adj. Cron α	Adj. Cron. α	Non adj. weigh kappa	Adj. weigh kappa	Non adj. Pear r (2-tailed)	Adj. Pear r (2-tailed)	Non adj. Cron α	Adj. Cron α
CORE COMPETENCIES												
1. Writing and editing	.352**	.526**	.577**	.74**	.673	.829	.44**	.752**	.455*	.866**	.625	.928
2. Document design	.311**	.478**	.513**	.692**	.678	.815	.525**	.614**	.644**	.765**	.748	.867
3. Rhetoric	.451**	.592**	.688**	.861**	.812	.919	.381**	.743**	.428*	.902**	.587	.942
4. Problem solving	.341**	.544**	.579**	.819**	.726	.893	.549**	.873**	.622**	.955**	.763	.977
5. Collaboration	.428**	.548**	.637**	.83**	.763	.902	.652**	.652**	.888**	.888**	.933	.933
6. Interpersonal communication	.232	.624**	.374**	.854**	.538	.912	.655**	.792**	.828**	.932**	.906	.965
7. Specialized expertise	.388**	.746**	.621**	.932**	.75	.961	.247*	.535**	.334	.774**	.489	.873
8. Technology	.299**	.562**	.522**	.857**	.659	.909	.451**	.559**	.617**	.776**	.762	.873
9. Overall score	.448**	.535**	.678**	.775**	.89	.861	.349**	.384**	.621**	.694**	.762	.812

*p < .05; **p < .01.

nonadjudicated Cronbach alpha analysis revealed that the nonadjudicated reader scores for writing and editing (α= .656), document design (α = .602), rhetoric (α = .656), collaboration (α = .631), specialized expertise (α = .691), and the overall e-portfolio score (α = .727) each achieved moderate to substantial correlations. Each score either remained the same or increased under adjudication.

Hence, it seemed reasonable to conclude that certain interreader correlations (among the variables of writing and editing, document design, rhetoric, collaboration, specialized expertise, and the overall score) had allowed us to reject the null hypothesis and accept the alternative hypothesis that there was, indeed, an acceptable level of interreader reliability. Based on reliable readings of these variables, we could then establish valid measures of central tendency, investigate interactions among the variables, and investigate each of their relationships to a criterion variable.

However, we could warrant no such evidence for the problem-solving variable, the interpersonal communication variable, and the technology variable. The weighted Kappa even for the adjudicated scores remained poor for the technology variable (K = .076, p < .05) and only slight for the problem-solving (K = .297) and the interpersonal communication variables (K = .111), neither of which met the 95% confidence interval. Analysis by means of Pearson's correlation coefficient revealed similar findings, with only the problem-solving variable reaching an acceptable level of agreement (r = .523, p < .05). Analysis by means of Cronbach's alpha revealed similar results, with only the problem-solving variable yielding an acceptable level of correlation (α = .68). While this variable may have been retained, the very low scores on the nonadjudicated scores, combined with the failure to meet the established confidence level in the weighted Kappa analysis, led us to remove these three variables from further analysis— such as documentation of measures of central tendency, core competency correlations, regression studies, and criterion variable estimation—in 2004.

In the spring of 2005, our instructors' scores revealed higher reliability coefficients, just as they had demonstrated greater agreement. While only three of the variables had met or exceeded the 95% confidence interval the previous year, now seven of the variables met that critical decision point, with four of the variables reaching the higher .01 confidence level: collaboration (K = .34), specialized expertise (K = .435), technology (K = .463), and the overall score (K = .53). Under adjudication, each of the variables reached the .01 confidence level, with moderate correlations above .4 for writing and editing, document design, rhetoric, problem solving, collaboration, interpersonal communication, technology, and the overall score. Specialized expertise now achieved a substantial correlation above .6. When Pearson's correlation was applied to the nonadjudicated scores in 2005, only the problem-solving score (r = .415) failed to meet the 95% confidence interval; when this measure was applied to the adjudicated scores, the range—from the correlation coefficient of the technology score ($r = .634$) to that of the specialized expertise score ($r = .858$)—was moderate to substantial, with each correlation coefficient meeting the higher .01 confidence interval. When examining the nonadjudicated scores under Cronbach's alpha, we found that the correlations were comparable to those achieved in the spring of 2004, with increases clear in problem solving ($\alpha = .819$), interpersonal communication ($\alpha = .876$), and technology ($\alpha = .767$). Hence, we felt justified in including each of the eight variables and the overall e-portfolio score for further analysis of the 2005 scores.

Warranting the inclusion of all variables has now become the norm of the program, as is depicted in Table 4.

During the 2006 and 2007 administrations of the assessment, only two instances were recorded of a level of agreement that failed to meet the 95% confidence interval: the nonadjudicated weighted Kappa recorded in 2006 for the interpersonal communication variable ($K = .232$) and the nonadjudicated Pearson correlation recorded in 2007 for the specialized expertise variable ($r = .334$). Remarkably, only the nonadjudicated weighted Kappa recorded in 2007 for the specialized expertise variable ($K = .247$) met the 95% confidence interval; each of the variables reached the more demanding .01 confidence interval. In 2006 and 2007, fair to moderate correlations are common in the nonadjudicated scores, and substantial to almost perfect correlation coefficients for the adjudicated scores are now frequent.

Ultimately, under conditions of maturation, the e-portfolios were able to be read reliably within our community—the first finding of the study. The scores demonstrated acceptable (and, in the majority of cases, substantial) levels of reader reliability, evidence of the ability of a community of readers to assess the core competencies of graduate-level technical communication. With issues of reliability resolved—a necessary precondition for further analysis—we now turn to our analysis of the scores themselves.

(2) What were the measures of central tendency?

The range, mean, and standard deviation scores are reported in Table 5 for 2004 through 2007.

As might be expected in any analysis of graduate student abilities, the scores are generally high. If we hold, as is traditional, that the midrange score is the combined score of seven (Elliot, Plata, & Zelhart, 1990)—one score from the upper range of the scale (4) and one score from the lower (3)—then the average scores are beyond this point for each of the core competencies and the overall e-portfolio score for 2004 and for all but one variable (interpersonal communication, $M = 6.32$, $SD = 2.7$) in 2005. There were no statistically significant differences in the mean scores across the first two administrations (writing and editing, $t(36) = -.090$, $p = .929$; document design, $t(36) = .415$, $p = .681$;

Table 5. Measures of Central Tendency, 2004 through 2007

YEAR	2004 (n varies from 13 to 17)			2005 (n = 22)			2006 (n = 31)			2007 (n = 27)		
MEASURE	Range	Mean	SD	Range	Mean	SD	Range	Mean	SD	Range	Mean	SD
CORE COMPETENCIES												
1. Writing and editing	7, 12	9.31	1.58	6, 12	9.36	1.84	6, 12	9.48	1.93	4, 12	9.52	1.93
2. Document design	7, 11	9.25	1.29	5, 12	9.00	2.14	5, 12	9.26	2.18	7, 12	9.33	1.69
3. Rhetoric	8, 12	9.41	1.22	7, 12	8.95	1.81	2, 12	8.83	2.26	2, 12	8.26	2.44
4. Problem solving	—[a]	—[a]	[a]	5, 11	8.45	1.87	2, 12	8.61	2.37	2, 12	7.26	3.10
5. Collaboration	6, 11	7.93	1.91	2, 11	8.00	2.26	2, 11	8.52	2.47	2, 12	7.96	2.85
6. Interpersonal communication	—[a]	—[a]	[a]	2, 12	6.32	2.7	2, 11	5.97	3.14	2, 12	6.11	3.45
7. Specialized expertise	7, 12	8.75	1.61	3, 12	8.05	2.42	2, 12	6.58	3.57	4, 12	8.52	2.17
8. Technology	—[a]	—[a]	[a]	6, 12	9.18	1.76	2, 12	9.13	2.77	4, 12	9.03	2.05
9. Overall portfolio score	7, 12	9.29	1.53	5, 12	8.95	2.01	4, 12	8.87	2.23	4, 11	8.29	2.01

[a]Variable not reliably read in 2004.

rhetoric, $t(37) = .893$, $p = .378$; collaboration, $t(35) = -.093$, $p = .926$; specialized expertise, $t(36) = 1.01$, $p = .319$; and overall score $t(37) = .578$, $p = .567$).

In 2006, as the assessment program matured, two variables were identified as below the score of 7: interpersonal communication ($M = 5.97$, $SD = 3.14$) and specialized expertise ($M = 6.58$, $SD = 3.57$). In 2007, the interpersonal communication variable remains below the score of 7, with no statistically significant difference between the mean scores from 2006 to 2007 ($t(56) = -.164$, $p = .867$). The ability of students to demonstrate interpersonal communication—a process-oriented, transactional construct—is difficult under any circumstance, even within an e-portfolio that is, after all, a product-oriented event. We have nevertheless retained the variable with the hope that podcasting technology, used in a new elective seminar in corporate communication, may allow students an opportunity to demonstrate presentation skills. A statistically significant difference between the mean score of the specialized expertise variable was identified from 2006 to 2007 ($t(56) = -.253$, $p < .05$), the improvement in 2007 an effort by the instructor of the communication theory and research seminar to make sure that each student had posted substantial evidence or work accomplished in that seminar in the e-portfolio.

The overall e-portfolio score—read holistically and, thus, perhaps the most robust indicator of student performance (Elliot, Briller, & Joshi, 2007)—remains solid for 2004 ($M = 9.29$, $SD = 1.53$), 2005 ($M = 8.95$, $SD = 2.01$), 2006 ($M = 8.95$, $SD = 2.23$), and 2007 ($M = 8.29$, $SD = 2.01$). As was the case in 2004 and 2005, there is no statistically significant difference in the overall mean e-portfolio scores from 2006 to 2007 ($t(56) = 1.023$, $p = .311$).

Thus, a second finding of the study may be documented: Students were able to meet or exceed our expectations on the majority of performance indicators as articulated in the core competencies and their NJIT descriptors. In terms of student performance, our students were meeting the program goals. Such outcomes suggest that the assessment plan presented here can be used in practice to assess student ability; thus, the performance-based assessment of graduate student skills in technical communication may be viewed as an empirically achievable fifth measure of the traditional model shown in Figure 1. To add strength to the potential of the new relational assessment model we were creating, we now turn to the internal consistency of the core competencies.

(3) What were the core competency interactions?

In order to investigate the interrelatedness of the core competencies, our variables, we performed the associative analysis shown in Tables 6 and 7.

Investigation of the spring 2004 assessment revealed statistically significant correlations between the overall e-portfolio score—taken in this analysis as the dependent variable in its relationship to the eight core competencies—and four of the independent variables: writing and editing ($r = .504$, $p < .05$), document

Table 6. Core Competency Correlations, 2004 through 2005

2004 (n varies from 13 to 17) / 2005 ($n = 22$)

	1. Writing and editing		2. Document design		3. Rhetoric		4. Problem solving		5. Collab.		6. Interper. Comm.		7. Specialized expertise		8. Technology		9. Overall score		10. Cum. GPA	
	04	05	04	05	04	05	04	05	04	05	04	05	04	05	04	05	04	05	04	05
1. Writing and editing	—	—	.646**	.714**	.507*	.848**	[a]	.794**	.121	.479*	[a]	.494*	.033	.649**	[a]	.742**	.504*	.79**	.423	.145
2. Document design	.646**	.741**	—	—	.759**	.713**	[a]	.595**	.374	.265	[a]	.578**	-.096	.7**	[a]	.733**	.888**	.753**	.415	.374
3. Rhetoric	.507*	.848**	.759**	.713**	—	—	[a]	.849**	.271	.44*	[a]	.51*	-.214	.75*	[a]	.673**	.895**	.757**	.326	.412
4. Problem solving	[a]	.794**	[a]	.595**	[a]	.849**	—	—	[a]	.73**	[a]	.499*	[a]	.69**	[a]	.768**	[a]	.854**	[a]	.446*
5. Collaboration	.121	.479*	.374	.265	.271	.44*	[a]	.73**	—	—	[a]	.296	.063	.304	[a]	.441*	.376	.574**	.268	.427*
6. Interpersonal communication	[a]	.494*	[a]	.578**	[a]	.51*	[a]	.499*	[a]	.296	—	—	[a]	.749**	[a]	.698**	[a]	.705**	[a]	.304
7. Specialized expertise	.033	.649**	-.096	.7**	-.214	.75*	[a]	.69**	.063	.304	[a]	.749**	—	—	[a]	.768**	-.079	.822**	-.171	.526*
8. Technology	[a]	.742**	[a]	.733**	[a]	.673**	[a]	.768**	[a]	.441*	[a]	.698**	[a]	.768**	—	—	[a]	.916**	[a]	.372
9. Overall score	.504*	.79**	.888**	.753**	.895**	.757**	[a]	.854**	.376	.574**	[a]	.705**	-.079	.822**	[a]	.916**	—	—	.370	.428*
10. Cumulative grade point average	.423	.145	.415	.374	.326	.412	[a]	.446*	.268	.427*	[a]	.304	-.171	.526*	[a]	.372	.370	.428*	—	—

[a] Variable not reliably read in spring 2004.

*$p < .05$; **$p < .01$.

Table 7. Core Competency Correlations, 2006 through 2007

2006 (n = 31) / 2007 (n = 27)

	1. Writing and editing		2. Document design		3. Rhetoric		4. Problem solving		5. Collab.		6. Interper. Comm.		7. Specialized expertise		8. Technology		9. Overall score		10. Cum. GPA	
	06	07	06	07	06	07	06	07	06	07	06	07	06	07	06	07	06	07	06	07
1. Writing and editing	—	—	.716**	.722**	.847**	.705**	.701**	.535**	.476**	.571**	.719**	.47**	.796**	.768**	.656**	.813**	.843**	.79**	.319	.322
2. Document design	.716**	.722**	—	—	.643**	.585**	.682**	.43*	.488**	.587**	.627**	.396**	.65**	.633*	.792**	.776**	.783**	.784**	.473**	.213
3. Rhetoric	.847**	.705**	.643**	.585**	—	—	.801**	.725**	.614**	.62*	.718**	.584**	.756**	.756**	.545**	.613**	.872**	.796**	.28	.548**
4. Problem solving	.701**	.535**	.682**	.43*	.801**	.725**	—	—	.794**	.622**	.646**	.835**	.716**	.6**	.569**	.512**	.799**	.65**	.207	.459*
5. Collaboration	.476**	.571**	.488**	.587**	.614**	.62*	.794**	.622**	—	—	.462*	.633**	.539*	.668**	.574**	.6**	.652**	.692**	.228	.218
6. Interpersonal communication	.719**	.47**	.627**	.396*	.718**	.584**	.646**	.835**	.462*	.633**	—	—	.72**	.422*	.504**	.374	.771**	.547**	.272	.202
7. Specialized expertise	.796**	.768**	.65**	.633**	.756**	.756**	.716**	.6**	.539*	.668**	.72**	.422**	—	—	.671**	.843**	.793**	.798**	.309	.458*
8. Technology	.656**	.813**	.792**	.776**	.545**	.613**	.569**	.512**	.574**	.6**	.504**	.374	.671**	.843**	—	—	.64**	.79**	.521**	.278
9. Overall score	.843**	.79**	.783**	.784**	.872**	.796**	.799**	.65**	.652**	.692**	.771**	.547**	.793**	.798**	.64**	.79**	—	—	.313	.423*
10. Cumulative grade point average	.319	.322	.473**	.213	.28	.548**	.207	.459*	.228	.218	.272	.202	.309	.458*	.521**	.278	.313	.423*	—	—

$*p < .05$; $**p < .01$.

design ($r = .888, p < .01$), and rhetoric ($r = .895, p < .01$). The writing and editing variable was correlated with both document design ($r = .646, p < .01$) and rhetoric ($r = .507, p < .05$). The document design variable was also correlated with rhetoric ($r = .759, p < .01$). No statistically significant correlations were observed between the collaboration or specialized expertise variables and any other variable.

A regression analysis of the spring 2004 e-portfolios relating the five reliably read independent variables (X) to the dependent variable, the overall e-portfolio score (Y), revealed an extraordinarily high coefficient of determination ($R^2 = .922, F(5, 9) = 23.226, p < .01$). That is, for the spring of 2004, 92% of the variability of the overall e-portfolio score represents the proportion of the variation in the dependent variable (the overall e-portfolio score) that is explained by the five independent variables (writing and editing, document design, rhetoric, collaboration, specialized expertise).

Investigation of the spring 2005 assessment revealed moderate to nearly perfect correlation coefficients that were statistically significant between the overall e-portfolio score and each of the eight core competencies: writing and editing ($r = .79, p < .01$), document design ($r = .753, p < .05$), rhetoric ($r = .757, p < .01$), problem solving ($r = .854, p < .01$), collaboration ($r = .574, p < .01$), interpersonal communication ($r = .705, p < .01$), specialized expertise ($r = .822. p < .01$), and technology ($r = .916, p < .01$). As was the case in 2004, the writing and editing variable remained correlated at statistically significant levels—yet now at a substantially higher level—with both document design ($r = .741, p < .01$) and rhetoric ($r = .848, p < .01$). The writing and editing variable was now shown to be correlated with each of the other variables as well: problem solving ($r = .794, p < .01$), collaboration ($r = .479, p < .05$), interpersonal communication ($r = .494, p < .05$), specialized expertise ($r = .649, p < .01$), and technology ($r = .742, p < .01$). While the document design variable remained correlated with rhetoric ($r = .713, p < .01$) and problem solving ($r = .595, p < .01$), that variable was now correlated with interpersonal communication ($r = .578, p < .01$), specialized expertise ($r = .7, p < .01$), and technology ($r = .733, p < .01$). The rhetoric variable was now correlated with problem solving ($r = .849, p < .01$), collaboration ($r = .44, p < .05$), interpersonal communication ($r = .51. p < .05$), specialized expertise ($r = .75. p < .05$), and technology ($r = .673, p < .01$).

While no correlation had been identified between the collaboration variable and any score or between the specialized expertise variable and any score in the spring of 2004, collaboration was now associated with writing and editing, as noted, as well as with problem solving ($r = .73, p < .01$) and with technology ($r = .441, p < .01$). Specialized expertise was correlated not only with writing and editing, document design, and rhetoric, as noted, and also with problem solving ($r = .69, p < .01$) and interpersonal communication ($r = .749, p < .01$). The three variables that were unable to be read reliably in 2004 were, in 2005, associated with other core competencies: problem solving was significantly associated with each of the other variables as well as with the overall e-portfolio score; interpersonal communication was significantly associated with each of the competencies, except collaboration; and technology was associated with each of the core competencies and the overall e-portfolio score. Indeed, in 2005, nearly all of the variables were significantly correlated at moderate to almost perfect levels with the exception of the following three: collaboration and document design (.265); collaboration and interpersonal communication (.296); and collaboration and specialized expertise (.304).

A regression analysis of the 2005 e-portfolios relating the eight independent variables that were reliably read to the overall e-portfolio score revealed, once again, an extraordinarily high coefficient of determination ($R^2 = .928$, $F(7, 14) = 25.86$, $p < .01$). That is, for the spring of 2005, 93% of the variability of the overall e-portfolio score represents the proportion of the variation in the overall e-portfolio score that is explained by the eight independent variables.

In 2006 and 2007, the patterns of association continue to be strengthened, as Table 7 demonstrates.

Overwhelmingly, there are moderate to nearly perfect correlation coefficients at the .01 level of statistical significance among the variables. Only a single correlation in 2007 (between interpersonal communication and technology) failed to meet the specified .05 confidence level. A regression analysis of the 2006 e-portfolios revealed, again, an extraordinarily high level of determination ($R^2 = .89$, $F(8, 22) = 22.23$, $p < .01$). Regression analysis of the 2007 data reveals that the same pattern is being maintained ($R^2 = .831$, $F(8, 18) = 11.04$, $p < .01$). Thus, our 4-year experiment has demonstrated that between 83% and 93% of the variability of the overall e-portfolio score represents the proportion of the variation in the overall e-portfolio score that is explained by the eight independent variables. The model thus appears to be both stable and robust.

A third finding of the study may thus be recorded: The model delineated in Figure 2 has yielded excellent levels of internal consistency. This feature demonstrates an empirical validity—scientifically oriented and model based—that is impossible to capture in the traditional audit model shown in Figure 1.

(4) What was the relationship of the core competencies to a criterion variable external to the present assessment process?

Significant to all writing assessment research must be the relationship of the core competencies expressed in the assessment model to independent, performance-based evidence. That is, if we were to make criterion-referenced interpretations for our stakeholders, further investigation into the model had to be made in order to analyze its association, if any, with that most readily available indicator of student performance: cumulative grade point average (GPA).

As Table 6 demonstrated, in the spring of 2004, no statistically significant correlations were identified between the six core competencies and the overall e-portfolio score with the cumulative GPAs of students in the sample. In the spring of 2005, however, analysis revealed correlations between the students' cumulative GPAs and problem solving ($r = .446, p < .05$), collaboration ($r = .427$, $p < .05$), specialized expertise ($r = .526, p < .05$), and the overall e-portfolio score ($r = .428, p < .05$). Table 7 shows statistically significant correlations in 2006 between the students' cumulative GPAs and only two variables: document design ($r = .473, p < .01$) and technology ($r = .521, p < .01$). In 2007, statistically significant correlations were observed between the students' cumulative GPAs and four variables: rhetoric ($r = .548, p < .01$), problem solving ($r = .459, p < .05$), specialized expertise ($r = .458, p < .01$), and the overall e-portfolio score ($r = .423, p < .05$). Such statistically significant correlations are encouraging, yet nevertheless serve as reminders that assessment of the e-portfolios, while a very high resolution snapshot, is only that. Powerful as the model is, a regression analysis of the 2007 data did not reveal a statistically significant correlation when the students' cumulative GPAs were taken as the dependent variable and the eight variables, along with the overall e-portfolio score, were taken as the independent variables ($R^2 = .453, F(9, 17) = 1.564, p = .2$).

Robust as the assessment model is, there are other factors that contribute to student achievement that must be captured by further investigation into the factors explored in this study. As a product, the e-portfolio by its very nature does not capture process-based student performance characteristics such as resilience (Sternberg & Subotnik, 2006) and intellectual style (Zhang & Sternberg, 2006). The ability of a student to prevail in a difficult seminar or a shift to a more effective intellectual style while engaging complex assignments would presumably influence an assigned grade. Thus, process-oriented cognitive complexity models (Torrance, van Waes, & Galbraith, 2007) hold the promise to reveal more about the effective performance of technical communication students as those performances are demonstrated within individual seminars.

Regardless of future study designs, however, key to the lack of statistically significant correlation coefficients between e-portfolio scores and the cumulative GPAs of our students is the extreme homogeneity of those GPAs in 2004 ($M = 3.91, SD = .148$), 2005 ($M = 3.87, SD = .205$), 2006 ($M = 3.87, SD = .289$), and 2007 ($M = 3.84, SD = .301$). Across all 4 years of the study, no statistically significant difference was identified among all recorded GPAs ($F(3, 93) = .243$, $p = .866$). While the emerging correlations seen in 2007 are somewhat promising, it is nevertheless true that scores on the core competencies did somewhat vary across the 4 years of the study; GPA did not. We had, then, told each other two different stories: the e-portfolio scores (a measure of student ability, highly reliable and apparently valid) varied; but the GPA (a measure of student ability with no reliability and unknown validity) did not. Only one of the two narratives can be correct, and it is our deep suspicion that the criterion measure in our study—the cumulative GPAs of our students—may indeed prove to be an insufficient measure of student achievement. As a measure of ability, GPA may underrepresent the construct of technical communication.

A fourth finding of the study was recorded: When performance assessment is in play, traditional criterion variables such as student cumulative GPA are very useful as heuristics to enable instructors to better understand the nature of their research findings. Programmatically, such a finding raises questions about the GPA itself in graduate programs such as ours. Because the e-portfolios are assessed during a limited time period, they are not as robust as observations made daily by individual instructors regarding student ability. On the other hand, it is naïve to assume that the e-portfolios, a mainstay of our program, should have no relationship to the grade point average. The answer, we have found, is not to embrace the value dualism implicit in the choice of validating either the e-portfolio or the cumulative GPA. Rather, we have come to be informed by the reliability of our e-portfolio readings and the strength of our model. With such consistency in mind, we have begun to rethink the singularity of our individual observations, valid as they may be. The truth is not to be hunted down either in singular class observations rendered longitudinally or in group observations voiced on a single day. A truth will be found in the emerging community that will enable us to reinvision our classroom observations even as we refine our group assessment. If programmatic alignment between e-portfolio and cumulative GPA eventually occurs, it will be because of community inquiry.

CONCLUSION: TRANSFER OF THE RELATIONAL ASSESSMENT MODEL

We began our investigation by seeking to improve on a received, traditional model of program evaluation. We end it by stating that the received model can be radically transformed by an insistence on a relational model. We now present that model in Figure 3.

The addition of the student e-portfolio shown within Figure 3—in this case, that of Carol Servino (2007), a program graduate who maintains her digital portfolio to continue to showcase her considerable talents—allowed us to refocus efforts. In the process of reconsidering the basic model offered in Figure 1, we fully operationalized the variables identified in Figure 2 to assess the e-portfolios of Ms. Servino and her classmates. We were able to conclude that that the e-portfolios were able to be read reliably, that students were able to meet or exceed our performance expectations, and that the core competencies, reliably measured, are well associated with each other and with the overall e-portfolio score. We even wondered publicly if the grade-point averages of our graduate students may be an insufficient measure of achievement. As Figure 3 demonstrates, an insistence on a relational model allows program developers to recast a static system into a dynamic context of relational information. The addition of a fifth measure to Figure 1 was not, therefore, additive, but

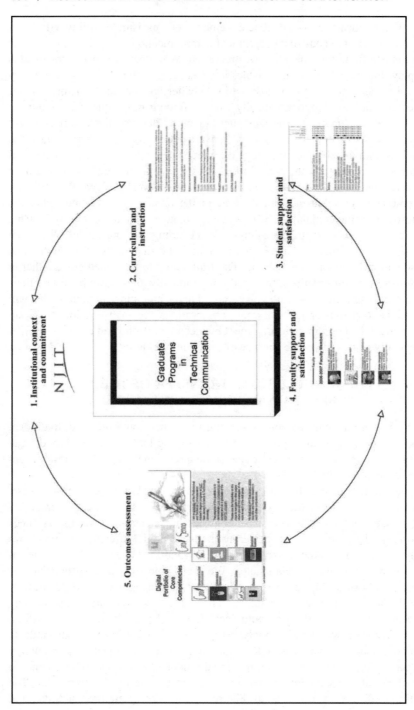

Figure 3. Relational assessment model for graduate programs in technical communication.

transformative, a strategy that allowed us to design an accountability model in which validation and values were seen as one (Kane, 2006; Messick, 1989), in which the stakeholder concerns of our students were recognized as significant (Guba & Lincoln, 1989), and where local meaning for ourselves and our students yielded context-sensitive and accessible information (Huot, 2002). With each passing semester, we continue to learn more about the value of valid performance assessment in program evaluation.

But, while we have articulated meaningful variables for the assessment of graduate students in technical communication, much more remains to be done. For example, our use of the weighted Kappa, Pearson correlation, and Cronbach's alpha are quite useful in determining the reliability of the readers, but these measures of reliability do not include the variables of the students themselves, do not treat the students as a random effect in the way suggested by Cherry and Meyer (1993, pp. 131–132). Are the reliability measures high enough? As Cherry and Meyer note, .94 is not realistic and .5 should leave us uncomfortable about making decisions about individuals; yet even their studied analysis of reliability does not provide a standard (p. 136). Further, should we be taking into account reader severity, as suggested by Nicholas T. Longford (1994) in his model of score adjustment? As well, there are consistent concerns with sample size in our study. As is commonly known, sample sizes under 30 are to be avoided with parametric investigations because the normal probability distribution cannot be achieved. And, finally, nagging questions of construct validity remain regarding the match between our assessment instrument and the lives lived in the asynchronous world of our classes. Have we, truly, captured the construct of graduate-level technical communication within our program? Questions concerning measurement and sampling plans and construct validity may be with us all days, but, together, we can offer a better set of answers.

Together, we can build a community for program assessment research, design new collaborative experiments in outcomes assessment, and universalize our model through collective action. We have a sustained history of research collaboration on outcomes assessment that began in 1988 with e-portfolio evaluation in the technical communication service course at NJIT (Elliot, Kilduff, & Lynch, 1994). Continuing in an assessment tradition designed to allow maximum construct representation, Coppola (1999) described a case study that demonstrated reliability, stability, and validity in its technical communication service-course assessment, tasks, and instructor community. At the present writing, this tradition of assessment has been maintained on the undergraduate level (Johnson, 2006), a testament to the transferability of the model. At the graduate level, a series of research presentations on program assessment (Coppola, 2005, 2006; Coppola & Elliot, 2003, 2004, 2006, 2007; Coppola, Elliot, Barker, Carter, Kimball, 2007) culminated in a 2006 grant from the Council for Programs in Technical and Scientific Communication (CPTSC), "A Communication Research Model for Assessment of Programs in Technical and Scientific Communication."

The model was offered to prompt—not to define—the development of universal, non–context specific programmatic elements that could be field-tested, modified, and validated by other graduate programs in the CPTSC community. We have already met the project goal of creating a Web-based forum for assessment research to view activities at multiple institutions. The CPTSC Assessment Research Model includes objectives of assessment practices, new assessment techniques used at various institutions, samples of student e-portfolios as they relate to the assessment techniques and data sets of assessment findings that may be used to foster collaborative studies. A second goal to perform at least one cross-institutional assessment in which outcomes data are generated, analyzed, and reported has also been met. In spring 2006, we partnered with the Master of Arts in Technical Communication (MATC) team at Texas Tech to share assessment strategies. The field test, described in Coppola and Elliot (2007), produced a collegial exchange and a combined set of core competencies—again, evidence of the transferability of the model described in this chapter.

As a community, there is much that our profession can accomplish. We may overcome the intellectual and financial drain on resources that will be prohibitive if each program in professional, scientific, and technical communication struggles to develop localized practices alone. If we continue to test the model across programs, we may work toward a coherent approach to a shared body of research and scholarship in our field. Johnson-Eilola and Selber (2004) acknowledge a disjunction in knowledge bases across the highly visible academic/practitioner divide as well as among various academic programs themselves: "Our field will not achieve the status of a mature profession until it can come to grips with a coherent body of disciplinary knowledge" (p. xxvii). If we work together to discuss, debate, and present assumptions of core competencies that drive assessment, we may develop universal, programmatic measures that can be tested and validated by graduate programs within specific institutional contents. We invite you to join us as we take these first steps.

ACKNOWLEDGMENTS

The authors acknowledge the contributions of portfolio readers Chris Funkhouser, Blake Haggerty, Carol Johnson, Burt Kimmelman, Robert Lynch, Robert Myre, and Kenneth Ronkowitz, as well as colleague Thomas Barker, Texas Tech University. The authors also thank Kamal Joshi, Database Manager in the NJIT Office of Institutional Research and Planning, for his help with the reliability analysis using the weighted Kappa statistic.

REFERENCES

Accreditation Board for Engineering and Technology (ABET). (2007). Retrieved November 4, 2007, from http://www.abet.org/

Allen, J. (1993). The role(s) of assessment in technical communication: A review of the literature. *Technical Communication Quarterly, 2*, 365–388.

Alred, G. (2003). Essential works on technical communication. *Technical Communication, 50*, 585–616.

American Educational Research Association (AERA), American Psychological Association (APA), & National Council on Measurement in Education (NCME). (1999). *Standards for educational and psychological testing.* Washington, DC: American Psychological Association.

Bernard, H. R. (2000). *Social research methods.* Thousand Oakes, CA: Sage.

Broadhead, G. J., & Freed, R. C. (1986). *The variables of composition: Process and product in a business setting.* Carbondale, IL: Southern Illinois University Press.

Certificate in the Practice of Technical Communication. (2007). Retrieved November 4, 2007, from http://adultlearner.njit.edu/programs/technicalcommunications-cert.php

Cherry, R. D., & Meyer, P. R. (1993). Reliability issues in holistic assessment. In M. H. Williamson & B. A. Huot (Eds.), *Validating holistic scoring for writing assessment: Theoretical and empirical foundations* (pp. 109–141). Creskill, NJ: Hampton Press.

Coppola, N. W. (1999). Setting the discourse community: Tasks and assessment for the new technical communication service course. *Technical Communication Quarterly, 8*, 249–267.

Coppola, N. W. (2005, March). *Writing assessment: A heuristic for graduate student success.* Paper presented at the 56th annual Conference on College Composition and Communication, San Francisco.

Coppola, N. W. (2006, March). Assessing graduate writing in a visual age: Towards core competencies. In N. Elliot (Session Chair). *In search of meaning: A community model for program assessment.* Paper presented at the 57th Annual Conference on College Composition and Communication, Chicago.

Coppola, N. W., & Elliot, N. (2003, October). *A behavioral framework for assessing graduate technical communication programs.* Paper presented at the Council for Programs in Technical and Scientific Communication. Abstract retrieved November 4, 2007, from http://www.cptsc.org/proceedings/2003/program_panels.htm

Coppola, N. W., & Elliot, N. (2004, October). *Towards formative assessment: Valuing different voices.* Paper presented at the Council for Programs in Technical and Scientific Communication. Abstract retrieved November 4, 2007, from http://www.cptsc.org/proceedings/2004/proceedings04_rev.pdf

Coppola, N. W., & Elliot, N. (2006, October). *A community research model for assessment of programs in technical and scientific communication.* Paper presented at the annual meeting of the Council for Programs in Technical and Scientific Communication, San Francisco State University.

Coppola, N. W., & Elliot, N. (2007). A technology transfer model for program assessment in technical communication. In K. St. Amant & C. Nahrwold (Eds.), Program assessment in technical communication [Special issue]. *Technical Communication, 54*, 459–474.

Coppola, N. W., Elliot, N., Barker, T., Carter, L., & Kimball, M. (2007, March). *Re(building) technical communication as a research discipline: A community model for program assessment.* Paper presented at the annual meeting of the Association of Teachers of Technical Writing, New York.

Council for Programs in Technical and Scientific Communication (CPTSC). (2007). Retrieved November 4, 2007, from http://www.cptsc.org/

Davis, M., Ramey, J., Williams, J., Gurak, L., Krull, R., & Steelhouder, M. (2003, September). *Shaping the profession: Leading academic programs in technical communication.* Paper presented at the meeting of the International Professional Communication Conference, Coronado Springs Resort. Orlando, FL.

Day, C. (2003, November 15–19). *Core competencies: A roadmap for action.* Paper presented at the 131st Annual American Public Health Association Meeting. Retrieved November 4, 2007, from
http://www.phf.org/Link/December03/Competencies_Abstract.pdf

Dayton, D., & Bernhardt, S. A. (2004). Results of a survey of ATTW members, 2003. *Technical Communication Quarterly, 13*, 13–43.

Dayton, D., Davis, M., Harner, S., Hart, H., Mueller, P., & Wagner, E. (2007, September). *Defining a body of knowledge.* Society for Technical Communication Academic-Industry Leaders Summit. University of Houston, TX.

Elliot, N. (2005). *On a scale: A social history of writing assessment in America.* New York: Peter Lang.

Elliot, N., Briller, V., & Joshi, K. (2007). Portfolio assessment: Quantification and community. *Journal of Writing Assessment, 3*, 5–30.

Elliot, N., Kilduff, M., & Lynch, R. (1994). The assessment of technical writing: A case study. *Journal of Technical Writing and Communication, 24*, 19–36.

Elliot, N., Plata, M., & Zelhart, P. (1990). *A program development handbook for the holistic assessment of writing.* Baltimore: University Press of America.

Elbow, P. (2006). Do we need a single standard of value for institutional assessment? An essay response to Asao Inoue's "community-based assessment pedagogy." *Assessing Writing, 11*, 81–99.

Fleckenstein, K. S. (2004). Words made flesh: Fusing imagery and language in a polymorphic literacy. *College English, 66*, 612–631.

Graham, S., & Perin, D. (2007). A meta-analysis of writing instruction for elementary students. *Journal of Educational Psychology, 9*, 445–476.

Guba, E. G., & Lincoln, Y. S. (1989). *Fourth generation evaluation.* Newbury Park, CA: Sage.

Haertel, E. H. (2006). Reliability. In R. L. Brennan (Ed.), *Educational measurement* (4th ed., pp. 65–110). Westport, CT: American Council on Education and Praeger.

Hamp-Lyons, L., & Condon, W. (2000). *Assessing the portfolio: Principles for practice, theory, and research.* Creskill, NJ: Hampton Press.

Harner, S. W., Johnson, R. R., Rainey, K. T., & Rude, C. (2003, October 3). Plenary panel of the 30th Annual Meeting of the Council for Programs in Technical and Scientific Communication, Clarkson University, Potsdam, NY.

Haswell, R. H. (Ed.). (2001). *Beyond outcomes: Assessment and instruction within a university writing program.* Westport, CT: Ablex.

Hollis, M. (1994). *The philosophy of social science: An introduction.* Cambridge, MA: Cambridge University Press.

Huot, B. (2002). *(Re)articulating writing assessment for teaching and learning.* Logan: Utah State University Press.

Inoue, A. B. (2005). Community-based assessment pedagogy. *Assessing Writing, 9*, 208–238.

Johnson, C. (2006). A decade of research: Assessing change in the technical communication classroom using online portfolios. *Journal of Technical Writing and Communication, 36*, 413–431.

Johnson-Eilola, J., & Selber, S. A. (Eds.). (2004). *Central works in technical communication.* New York: Oxford University Press.

Kane, M. T. (2006). Validation. In R. L. Brennan (Ed.), *Educational measurement* (4th ed., pp 17–64). Westport, CT: American Council on Education and Praeger.

Landis, J. R., & Koch, G. G. (1977). The measurement of observer agreement for categorical data. *Biometrics, 33*, 159–174.

Longford, N. T. (1994). Reliability of essay rating and score adjustment. *Journal of Educational and Behavioral Statistics, 19*, 171–200.

Master of Science in Professional and Technical Communication (MSPTC). (2007). Retrieved November 4, 2007, from http://msptc.njit.edu/

Messick, S. (1989). Validity. In R. L. Linn (Ed.), *Educational measurement* (3rd ed., pp. 13–103). New York: American Council on Education and Macmillan.

Messick, S. (1994). The interplay of evidence and consequences in the validation of performance assessments. *Educational Researcher, 23*, 13–23.

Middle States Commission on Higher Education (MSCHE). (2007). Retrieved November 4, 2007, from http://www.msche.org/

New Jersey Institute of Technology. (2007). Retrieved November 4, 2007, from http://www.njit.edu/

Purvis, A. C., Gorman, T. P., & Takala, S. (1988). The development of the scoring scheme and scales. In T. P. Gorman, A. C. Purves, & R. E. Degenhart (Eds.), *The IEA study of written composition I: The international writing tasks and scoring scales* (pp. 41–58). Oxford: Pergamon Press.

Servino, C. (2007). *Digital portfolio of core competencies.* Retrieved November 4, 2007, from http://web.njit.edu/~cs222/

Spearman, C. (1904a). The proof and measurement of association between two things. *American Journal of Psychology, 15*, 72–101.

Spearman, C. (1904b). 'General intelligence,' objectively determined and measured. *American Journal of Psychology, 15*, 201–293.

Stemler, S. E. (2004). A comparison of consensus, consistency, and measurement approaches to estimating interrater reliability. *Practical Assessment, Research & Evaluation, 9*. Retrieved November 4, 2007, from http://PAREonline.net/getvn.asp?v=9&n=4

Sternberg, R., & Subotnik, R. F. (2006). *Optimizing student success in school with the other three Rs: Reasoning, resilience, and responsibility.* Greenwich, CT: Information Age Publishing.

Tinder, G. (1980). *Community: Reflections on a tragic ideal.* Baton Rouge, LA: Louisiana State University Press.

Torrance, M., van Waes, L., & Galbraith, D. (Eds.). (2007). *Writing and cognition: Research and applications.* Amsterdam, The Netherlands: Elsevier.

White, E. M. (1994). *Teaching and assessing writing* (2nd ed.). San Francisco, CA and London: Jossey-Bass.

Wiggins, G. W. (1993). *Assessing student performance: Exploring the purpose and limits of testing.* San Francisco, CA and London: Jossey-Bass.

Wiggins, G. W. (1994). The constant danger of sacrificing validity to reliability: Making writing assessment serve writers. *Assessing Writing 1*, 129–139.

Witte, S. P., & Faigley, L. (1983). *Evaluating college writing programs*. Carbondale and Edwardsville, IL: Southern Illinois University Press.

Yancey, K. B. (2004). Postmodernism, palimpsest, and portfolios: Theoretical issues in the representation of student work. *College Composition and Communication, 55*, 738–761.

Zhang, L., & Sternberg, R. J. (2006) *The nature of intellectual styles*. Hillsdale, NJ: Erlbaum.

CHAPTER 10

Program Assessment, Strategic Modernism, and Professionalization Politics: Complicating Coppola and Elliot's "Relational Model"

Gerald Savage, Illinois State University

Nancy Coppola and Norbert Elliot (see Chapter 9, this volume) implicitly acknowledge that they are working in a world in which the tapestry of science, very much like Penelope's tapestry in Homer's *Odyssey*, is continually and strategically woven and unwoven from discursive and political threads. Initially seeming to undertake a blandly mainstream program assessment that adheres to a classical empiricist paradigm for statistical reliability, Coppola and Elliot quickly and usefully complicate the task by pointing out that validity requires admitting complexity. The project they set for themselves is ambitious, sophisticated, and takes them well beyond what most of us would consider the program assessment comfort zone—a zone that is, at its best, not a place anyone seeking comfort would retreat to. Program assessment is an institutional necessity, a fact we will probably acknowledge as having value, even though we seldom welcome it because of its unavoidable imposition on faculty and staff, whose commitments already take more time than having a life beyond the office can accommodate. But the assessment process has relatively standard features and methods that are easy enough to understand.

We grumble that such assessments miss much—perhaps most—of what we and our students do best and value most in our programs, but as Frost's grindstone-cranker philosophizes, "What if it wasn't all it should be? I'd be satisfied if he'd be satisfied" (p. 232). But this was not the philosophy of Coppola, Elliot, and their colleagues as they prepared to do their 2004 program assessment. In reviewing the results of their previous assessment, they note, Scripted as the review process itself may be, the traditional auditing model nevertheless has its virtues (p. 128). It provided data regarding institutional support, student satisfaction, curricular success in relationship to benchmark programs, and faculty satisfaction. They were able to evaluate these factors in relation to program goals and to produce what seems to have been a detailed and comprehensive overview of their program's accomplishments and to set goals for further improvement and growth. Nevertheless, they concluded that "the model had its limits" (p. 130), that despite helping them to define their four "predictor variables," it did not enable them to determine whether they were meeting program objectives.

Very much to their credit, they "decided that it was time to do better." Specifically lacking, they believed, was an assessment of student performance. Adding a single variable seems like a modest step toward increased complexity, but it proved to be a step into unknown territory because of the lack of prior research on assessment of graduate programs in technical communication. Moreover, because technical communication has, as yet, few if any established standards for disciplinary knowledge, or what is coming to be called "core competencies," they drew on work by college-level composition scholars Kathleen Blake Yancey, Stephen Witte and Lester Faigley, and others, adapting those perspectives to technical communication.

Core competencies were identified on the basis of two studies in technical and professional communication: one a survey of core competencies by David Dayton and Stephen Bernhardt and the other an annotated bibliography of "essential works" in the field compiled by Gerald Alred. Coppola and Elliot developed a list of eight traits from these studies to use in competency assessment. Their use of these studies in the formulation of assessment criteria is perhaps obvious in retrospect but profoundly significant as a strategic use of new research, having important political implications for the technical communication field. Much of the remainder of this essay will address these implications.

It is a significant development for our field that research has turned toward exploration and theorizing of disciplinary standards and competencies, but Dayton and Bernhardt's study and Alred's bibliography have been both preceded and followed by several significant projects by other scholars, which only marginally overlap with the ones Coppola and Elliot use. Notable examples can be found in Hart-Davidson and in Wick; and especially in numerous studies by Saul Carliner. Additionally, bibliographic studies by Elizabeth Overman Smith and recent collections of essays considered by their editors to be standard or groundbreaking works in technical communication, argue for the emergence of a canon of technical communication "literature" (Dubinsky, 2004; Johnson-Eilola & Selber, 2004; Peeples, 2003).

I do not suggest that Coppola and Elliot are unaware of the rapid evolution of knowledge in the field—quite clearly they are not; nor do I suggest that they are misguided in employing what might be characterized as preliminary research on core competencies as a basis for defining assessment criteria. Instead, what I see Coppola and Elliot doing is strategic work that has implications not just for their own program's credibility and status at NJIT but for the field as a whole. As well, I see their work in at least two different perspectives: strategic in its employment of a modernist assessment paradigm that will be persuasive to university administrators and program-assessment teams; and political in its effort to move the field of technical communication toward disciplinary standards that will give us greater status as a profession. In the following two sections, I will elaborate on these perceptions.

STRATEGIC MODERNISM IN ASSESSMENT METHODOLOGY

Coppola and Elliot's study is rigorously and conservatively designed and conducted. Its methodology and rationale are clearly set forth. Although my eyes quickly glaze when they fall upon terms such as Cronbach's Alpha, Pearson's R, One-tailed value, and most other statistical terminology, the tables and graphs are generally understandable and the findings not too difficult to correlate with the methodology. Moreover, for those who are not put off by research conclusions that are consciously ambiguous (I am generally among those not so put off), the study provides us much to think about. The ambiguity of the study may be most obvious in the purpose statement and the conclusion. The purpose, "to offer a consolidated model for the assessment of graduate programs in technical writing" (p. 127), is contradicted by the (penultimate) conclusion, that "much more remains to be done" (p. 155).

As Coppola and Elliot point out, empirical data in the social sciences—many would include the "natural sciences" as well—is always already uncertain, contingent, and unavoidably freighted with values, motives, and points of view. They express this point in epigrammatic language, pointing out that causal certainty will always evade us. Their real conclusion—that the standards our field needs cannot be established empirically but only through political processes of communal dialog and negotiation—situates this study, conducted in the best modernist tradition, within the postmodern dilemma.

Yet, as program directors who must spend our professional lives in modernist institutions, we must argue for the worth of our curricula, our teaching, and the learning of our students in measurable terms, using criteria that make sense to our profession as a whole, as well as to the professions our field interacts with, and to the markets that purchase our professional services and products. These internal and external arguments and strategies, such as those I suggest are evident in Coppola and Elliot's study, are characteristic of processes going on at various levels and on various scales throughout our field. Together, these processes are part of the politics of professionalization.

STANDARDS, COMPETENCIES, AND THE POLITICS OF PROFESSIONALIZATION

The challenges of Coppola and Elliot's assessment project intersects with a broader challenge for our field—the struggle for professional status—which is complicated further by the question of the future of the professions as knowledge itself becomes less a foundation than a commodity in the economy of the information society (Bernhardt & Farmer, 1998; Faber & Johnson-Eilola, 2002, 2003). As I and others have argued in recent years, the field of technical communication has not yet attained the status of a mature profession (Davis, 2001; Killingsworth, 1999; Pinelli & Barclay, 1992; Savage, 1999). The social and economic processes leading to a field's achieving professional status have been studied and theorized extensively by scholars in the sociology and history of professionalization (see, for example, Burrage & Torstendahl, 1990; Freidson, 1986; Haskell, 1984; Torstendahl & Burrage, 1990). Such studies indicate that professions emerge in processes of struggle for market control and closure, for definition of a coherent body of knowledge, and for development of a professional history that will give the field a unifying identity, among other factors. Once we understand something about these processes, they become apparent in many of the activities in our own field.

We see evidence of where we are in the professionalization struggle in Coppola and Elliot's observation that technical communication as a field does not yet have an agreed-upon body of professional knowledge, which in an applied field will necessarily be manifested as core competencies. These issues take on a new level of concern for those of us charged with development of assessment criteria. Assessment processes presuppose a standard body of knowledge for the field in which an academic program participates. Indeed, it can be argued that universities presuppose such a status for the knowledge represented in any program they house. Without a defined body of knowledge, it is impossible to do meaningful program assessment, which puts at risk the entire agenda of achieving disciplinary status and recognition.

The question of certification arises in connection with the difficulties Coppola and Elliot encountered, as well. Certification—as advocates such as Carliner (1996), Freeman and Spyridakis (2004), Rude (2004), Hayhoe (2000), and Turner and Rainey (2004) have argued—is necessary for full professional status, and itself requires a standard body of knowledge as a basis for certification decisions. Arriving at these destinations, however, is not necessarily a linear process. We are confounded by the problem we frequently hear about: that our field has no body of knowledge of its own because we have no useful research of our own to produce such knowledge (see Pinelli & Barclay, 1992, among others). Various blame has been laid for these difficulties, but surely part of the difficulty is that lacking professional status, standardized knowledge, and standard practices, there is no clearly defined professional practice on which to focus any research we might want to conduct.

Many of these difficulties seem to have identifiable causes and concrete definitions. But what is missing from many of these discussions is the aspect of power, and not simply the lack of power that is to be expected for a field that is not recognized or understood. What may be more pertinent to our current situation is how movement toward professionalization occurs. Such moves are inevitably political, of course. But political processes are often interpreted as exercises of power, whereas in everyday life, politics often proceeds almost invisibly in processes of ideological change, changes in values and beliefs, or what David Forgacs (1993) describes as "a coming to consciousness of a coincidence of interests." The work of Italian political philosopher and anti-Fascist activist Antonio Gramsci (1971) is particularly useful in understanding such processes, and although he was concerned with national political issues, his work has come to be applied in a number of areas, including critical theory, cultural studies, literacy, and writing studies.

A study such as the one described by Coppola and Elliot employs a methodology that was validated by NJIT for purposes that are valued by NJIT. I do not suggest that the study was done as a political tactic within a larger project of professionalization, but professionalization tends to proceed very much in the Gramscian sense of conceptions of the world realized in political action (Gramsci, 1971, pp. 326–327); that is, through moves with immediate, practical goals for which professional status may not even be on anyone's mind. Yet it accomplishes certain professionalizing functions. Simply by designing a study aimed at developing a model for program assessment, the authors demonstrate responsibility toward the norms of the academy and the broader social community. Such displays of responsibility and "normality" accrue credibility— belief—not only for the NJIT program in professional and technical communication but for the field as a whole. By selecting Bernhardt and Dayton's survey data and Alred's list of essential works as their sources of core competencies and field knowledge, they make a move toward canonization of those concepts. Again, I do not suggest that the choice of these resources was overtly or explicitly political, but, in the Gramscian sense, by making the choice, their action is political. Publication of the study in this book with a prestigious press adds further credibility to the NJIT program, to their assessment criteria, to the implicit canonicity of the sources, and to the field as a whole.

Many of us are concerned about the potential consequences of assessment for our own programs, particularly in cases where we may have little voice in the process or the criteria applied. The proactive approach by Coppola and Elliot provides an example of a way to anticipate program reviews we might prefer to avoid. Indeed, the very exercise of systematically examining what, how, and why we do what we do in our programs in order to give an accounting to those who have the power to influence our programs for good or ill forces us to think about and seek convincing arguments for the big questions we might otherwise prefer never to consider: What is the core knowledge of our field? What distinguishes our field from others that may claim similar knowledge? How are the competencies of scholars, students, and practitioners in the field assessed, and how do our assessment methods compare with those used in other programs in the field?

The authors also refer, obliquely, to a concern for colonialism. Although they do not elaborate on this concern, it seems they may sense a dark side to the professionalization moves they are furthering. It is significant that the first task called for in Coppola and Elliot's vision of "a new community of researchers" is "establishing core competencies," an issue that has been on the minds of a number of scholars in our field in recent years, and another clear phase in the professionalization process. It is significant, as well, that in its present form, this call seems to conceptualize such a community in normative terms—note the use of such verbals as "establishing," "tailoring," and "publicizing," and the general emphasis on assessment, models, and performance.

The research community the authors have in mind is, of course, intended to serve programmatic goals, not the development of new knowledge in the field. It does, nevertheless, seem to raise the question of colonization that the authors had hoped to abandon. On the other hand, it suggests a postmodern solution to the postmodern dilemma concerning knowledge. It frames the notion of research in terms of power rather than of value-neutral "inquiry," and assigns to exploration the tasks of locating the most effective political tools, that is, tools for assessment and communication channels. I point to these things to acknowledge the choices it seems we have to make in order to maintain the purchase we so far have, which is a steady increase of legitimacy and autonomy within the academy, the "guarantor" of our specialized disciplinary knowledge. As Geisler argues, the privileges and status of professions are established on the "cognitive bedrock" of socially valued, expert knowledge (p. 72). The validity of such expertise is certified by the universities in which it is acquired. But professional autonomy is achieved in a process of struggle in relation to contending interests in the labor market. Thus, the formation of a professional community that will isolate and claim exclusive ownership of the competencies that define the profession is a political and colonizing process. In abandoning the notion of knowledge-making as an objective enterprise of inquiry, instead embracing knowledge as communally negotiated, we also inevitably surrender the innocence of classical science in exchange for political consciousness and the inevitability of interests in whatever actions we choose to take.

REFERENCES

Bernhardt, S. A., & Farmer, B. W. (1998). Work in transition: Trends and implications. In M. S. Garay & S. A. Bernhardt (Eds.), *Expanding literacies: English teaching and the new workplace* (pp. 55–80). Albany, NY: SUNY Press.

Burrage, M., & Torstendahl, R. (Eds.). (1990). *Professions in theory and history: Rethinking the study of the professions.* Newbury Park, CA: Sage.

Carliner, S. (1992). What you should get from a professionally oriented master's degree program in technical communication. *Technical Communication, 39*(2), 189–199.

Carliner, S. (1996). Evolution-revolution: Toward a strategic perception of technical communication. *Technical Communication, 43*(3), 266–276.

Carliner, S. (2000a). Intellectual capital: Placing a value on technical communication. *Intercom, 47*(8), 7–9.

Carliner, S. (2000b). Physical, cognitive, and affective: A three-part framework for information design. *Technical Communication, 47*(4), 561–576.

Carliner, S. (2000c). Trends for 2000: Thriving in the boom years. *Intercom, 47*(1), 11–14.

Carliner, S. (2001). Emerging skills in technical communication: The information designer's place in a new career path for technical communicators. *Technical Communication, 48*(2), 156–175.

Carliner, S. (2003). Characteristic-based, task-based, and results-based: Three value systems for assessing professionally produced technical communication products. *Technical Communication Quarterly, 12*(1), 83–100.

Davis, M. T. (2001). Shaping the future of our profession. *Technical Communication, 48*(2), 139–144.

Dubinsky, J. M. (Ed.). (2004). *Teaching technical communication: Critical issues for the classroom.* Boston, MA and New York: Bedford/St. Martin's.

Faber, B., & Johnson-Eilola, J. (2002). Migrations: Strategic thinking about the future(s) of technical communication. In B. Mirel & R. Spilka (Eds.), *Reshaping technical communication: New directions and challenges for the 21st century* (pp. 135–148). Mahwah, NJ: Lawrence Erlbaum Associates.

Faber, B., & Johnson-Eilola, J. (2003). Universities, corporate universities, and the new professionals: Professionalism and the knowledge economy. In T. Kynell-Hunt & G. J. Savage (Eds.), *Power and legitimacy in technical communication: The historical and contemporary struggle for professional status* (Vol. 1, pp. 209–234). Amityville, NY: Baywood.

Flannery, K. T. (1995). *The emperor's new clothes: Literature, literacy, and the ideology of style.* Pittsburgh, PA: University of Pittsburgh Press.

Forgacs, D. (1993). National-popular: Genealogy of a concept. In S. During (Ed.), *The cultural studies reader* (pp. 177–190). London: Routledge.

Freeman, K. S., & Spyridakis, J. H. (2004). An examination of factors that affect the credibility of online health information. *Technical Communication, 51*(2), 239–263.

Freidson, E. (1986). *Professional powers: A study of institutionalization of formal knowledge.* Chicago, IL: University of Chicago Press.

Freire, P., & Macedo, D. (1987). *Literacy: Reading the word and the world.* New York: Bergin and Garvey.

Frost, R. (1949). The grindstone. *Complete poems of Robert Frost* (pp. 232-234). New York: Holt, Rinehart and Winston.

Geisler, C. (1994). *Academic literacy and the nature of expertise: Reading, writing, and knowing in academic philosophy*. Hillsdale, NJ: Lawrence Erlbaum Associates.

Gramsci, A. (1971). *Selections from the prison notebooks of Antonio Gramsci* (Q. Hoare & G. N. Smith, Trans.). New York: International Publishers.

Grimm, N. M. (1996). Rearticulating the work of the writing center. *College Composition and Communication, 47*(4), 523–548.

Hart-Davidson, W. (2001). On writing, technical communication, and information technology: The core competencies of technical communication. *Technical Communication, 48*(2), 145–155.

Haskell, T. L. (Ed.). (1984). *The authority of experts: Studies in history and theory*. Bloomington, IN: Indiana University Press.

Hayhoe, G. F. (2000). What do technical communicators need to know? *Technical Communication, 47*(2), 151–153.

Johnson-Eilola, J., & Selber, S. A. (Eds.). (2004). *Central works in technical communication*. New York: Oxford University Press.

Killingsworth, M. J. (1999). Technical communication in the 21st century: Where are we going? *Technical Communication Quarterly, 8*(2), 165–174.

Peeples, T. (Ed.). (2003). *Professional writing and rhetoric: Readings from the field*. New York: Addison Wesley Educational Publishers.

Pinelli, T. E., & Barclay, R. O. (1992). Research in technical communication: Perspectives and thoughts on the process. *Technical Communication, 39*(4), 526–532.

Rude, C. (Ed.). (2004). Special issue: The state of technical communication in its academic context, part 2. *Technical Communication Quarterly, 13*(2).

Savage, G. J. (1999). The process and prospects for professionalizing technical communication. *Journal of Technical Writing and Communication, 29*(4), 355–381.

Smith, E. O. (2000). Strength in the technical communication journals and diversity in the serials cited. *Journal of Business and Technical Communication, 14*(2), 131–184.

Smith, E. O. (2004). Points of reference contributing to the professionalization of technical communication. In T. Kynell-Hunt & G. J. Savage (Eds.), *Power and legitimacy in technical communication: Strategies for professional status* (Vol. 2, pp. 51–72). Amityville, NY: Baywood.

Torstendahl, R., & Burrage, M. (Eds.). (1990). *The formation of professions: Knowledge, state and strategy*. Newbury Park, CA: Sage.

Trimbur, J. (1994). The politics of radical pedagogy: A plea for 'a dose of vulgar Marxism.' *College English, 56*(2), 194–206.

Turner, R. K., & Rainey, K. T. (2004). Certification in technical communication. *Technical Communication Quarterly, 13*(2), 211–234.

Wick, C. (2000). Knowledge management and leadership opportunities for technical communicators. *Technical Communication, 47*(4), 515–529.

Wright, E. (2003). Reading the cemetery, lieu de memoire par excellence. *Rhetoric Society Quarterly, 33*(2), 27–44.

Technology in Assessment

CHAPTER 11

Assessing Professional Writing Programs Using Technology as a Site of Praxis

Jeffrey Jablonski and Ed Nagelhout
University of Nevada Las Vegas

This chapter describes assessment as a primary tool for developing a professional writing program, with technology as the primary site of praxis. Professional writing programs in universities across the country have grown exponentially, and a strong program requires long-term planning and implementation strategies. To think programmatically is to understand that professional writing programs, as local constructs, cannot exist in a vacuum nor exist as the vision of a single individual. Professional writing programs are the property of all stakeholders, and a complex use of assessment strategies should underlie program development goals.

In building a comprehensive professional writing program, we assume a certain number of primary goals for the program: to offer consistent curricula, to enhance instruction, and to aid in the professional development of teachers in the program. To meet these goals, we recommend using technology as a means to incorporate a variety of assessment plans central to each phase of program development—planning, implementation, and completion (Eubanks & Abbott, 2003). In this chapter, we will explain the crucial role that assessment should play in the short-term and long-term development plans of professional writing programs. We describe these assessment plans in light of complex adaptive systems theory, participatory action research, and user-centered design; for us, these theoretical constructs offer the most effective ways to enact the reciprocity necessary between assessment and professional writing program design. We use our own experiences developing a University of Nevada Las Vegas (UNLV) professional writing "service" course into a coherent curricular unit as an example of how an assessment strategy centered in technology can enhance the overall development of a professional or technical writing program.

GOALS FOR THE PROFESSIONAL WRITING
PROGRAM AT UNLV

UNLV, classified by the Carnegie Foundation as Doctoral/Research University-Intensive, is a metropolitan university with a primarily nonresidential student population of over 27,000. Our professional writing program consists of the traditional upper-division "service" courses aimed at introducing majors across the university to the genres of technical and workplace communication. Our discussion will focus on our business writing course, which became an English department offering in 1996. We were originally hired to create and teach in the new program after the course was moved from the business college, based on a recommendation from their accrediting association. By 2003, our department was offering over 50 sections annually, taught in computer classrooms and in distance-education format.

Located in one of the fastest growing cities in the country, the university's enrollments are projected to increase 8% annually, with the student population predicted to reach 40,000 by the year 2010. Within this context, we feel strongly that the professional writing program needs to keep pace with this growth and develop a standardized curriculum that prepares students for communicating in today's information- and technology-rich environments. Because of the rapid rate at which information technologies change, our writing program also needs to become dynamic enough to incorporate the latest shifts in how people communicate in the workplace.

Following scholarship on writing program administration (e.g., Rose & Weiser, 1999), we assume a writing program is a complex curricular system of students and teachers. A consistent curriculum, to us, means that the program has a united vision for teaching and learning in the professional writing classroom. This includes a common syllabus, common readings, a core set of assignments, and, most importantly, a common set of goals and expectations that would lead to a shared set of outcomes. For the students, an important goal for us is that each one has a similar experience. We began with an assumption that students who are expected to write in technical and professional settings need to develop multiple literacies—rhetorical, visual, information, computer, and ethical (Nagelhout, 1999). A postmodern/postindustrial society demands that citizens acquire multiple literacies for interacting with more sophisticated information mediums. We want our students to become more information savvy, to understand disciplinary discourse conventions, and recognize the constructed nature of knowledge so that they can use knowledge effectively in discipline-specific situations. We want our students to become more flexible in their thinking (especially as active users of technology) by understanding the visual nature of writing, developing a sense of perceptual organization, and seeing page design and layout as integral components of successful documents. Finally, we believe that students need to make informed and ethical choices appropriate for particular rhetorical situations, yet also understand the implications and consequences of those choices.

Tied closely to these goals for the classroom is a belief that students need access to cutting-edge writing technologies and multiple spaces for interaction. Robert Kramer and Stephen Bernhardt's (1999) report on Glyph, a Web-based instruction environment at New Mexico State, showed how moving instruction to online spaces facilitates the development of rhetorical, visual, and "multi-tasking" (i.e., computer) literacy. A Web-based professional writing curriculum would provide accessible modes of instruction and resources that help guide students more effectively and provide them with the kinds of experiences necessary to participate in an information-rich society. In our program, we want students to understand electronic writing and communication technologies used in the workplace, and more importantly, understand the implications of their use; in other words, we want students to actively engage with technology rather than be passive users. In making our professional writing courses a positive learning experience, we have to acknowledge that our students have diverse learning styles; moreover, with the majority of our students living off campus and working from half-time to full-time, it is important that students have multiple spaces to interact with the instructor, with individuals, with small groups, and with the whole class. And we need this to occur in both face-to-face and electronic environments.

For teachers, we assume they need multiple forms of support to ensure quality instruction, and, as the Texas Tech Online-Print Integrated Curriculum (TOPIC) system has demonstrated, electronic technologies can facilitate their professionalization (Kemp, 2003; Salvo, 2001). We want to ensure a formal support system for teachers that is embedded as a critical component of the overall program structure. We want teachers to be comfortable in the classroom, comfortable with the curriculum, and comfortable with their professional positions within the university. To do this, we needed to create a professional atmosphere that includes mentoring, regular staff meetings, professional development opportunities, and a legitimate stake in the growth of the program.

Overall, we want a professional writing program that is responsive to the needs of its stakeholders. This kind of flexibility is imperative to our vision. But this kind of flexibility seemed incompatible with traditional methods of assessment and evaluation, nor could it rely on traditional methods for eliciting input from teachers and students.

PROFESSIONAL WRITING PROGRAM DEVELOPMENT AND PRAXIS-ORIENTED ASSESSMENT METHODS

Our thinking about assessment is motivated less by pressure from external audiences to demonstrate results and more by an "internal" desire to achieve the curricular coherence we described in the previous section. As Jo Allen explains, assessment should not be viewed as a co-option of our responsibilities or some mass "plot to expose our cushy academic lifestyle," but rather as "an opportunity to learn what we are doing right as educators and what we are doing wrong" (1993, p. 1). Planning assessment begins with understanding what should be assessed (typically, student learning, teaching effectiveness, or program effectiveness) and how the results will be used (to evaluate students, teachers, or programs) (Allen, 1993). Consequently, our focus is primarily on our efforts to assess the effectiveness of the program, rather than the students or the teachers. Moreover, since we are in the process of developing the program, we are interested in assessment as a means of gathering information that would allow us to determine if we are meeting our goals as outlined previously and what improvements are necessary.

While research on assessment of academic technical communication programs is burgeoning (see Hundleby, Hovde, & Allen, 2003), it is still primarily focused on assessing student performance, or the extent to which programmatic goals are observable in student performance (see Cook, 2003). In her description and rationale for portfolios as a method for assessing student performance, Nancy Coppola (1999) asserts that the iterative nature of determining a rubric—collecting portfolios, and conducting evaluations across several semesters—contributes to programmatic consistency over time. We agree with Coppola that assessment activity in general adds to curricular coherence; however, she was not directly assessing her program. Philip Eubanks and Christine Abbot (2003) report on the uses of the focus-group method as a means of assessing programs at three stages of development: the planning phase ("needs assessment"), the implementation phase ("formative assessment"), and the completion phase ("formative assessment"). Eubanks and Abbott give us basic concepts to situate stages of assessment, and one particular method of assessment, but not a framework for measuring the student-centered and teacher-centered program we envisioned.

Because we assume a writing program to be a complex structure consisting of many stakeholders, we turned to praxis-oriented methods: complex adaptive systems theory, user-centered design, and action research. Complex adaptive systems theory posits that healthy and self-producing systems have a drive from within to continue living (Marion, 1999; Rasch & Wolfe, 2000). Healthy systems are self-bounded, self-regenerating, and self-perpetuating. Obviously, we could not incorporate the theory wholesale into our planning, but we were particularly struck by three key concepts: autopoiesis, nonlinear dynamics, and feedback loops. Autopoiesis literally means "self-producing." For systems to be autopoietic—as defined by John Mingers in *Self-Producing Systems* (1995)—they must be "organized in such a way that their processes produce the very components necessary for the continuance of these processes" (p. 11). And while some would posit that self-producing organizations need to be more concerned with self-preservation, regardless of their external environment, the successful writing program cannot be so "egocentric." In this respect, our professional writing program (as a system) needed to establish a clear set of boundaries, such as clearly articulated goals and objectives for our courses. These boundaries, based in part on institutional resources and relationships, identify our program's responsibilities as a part of the larger university system.

While linear dynamics are those in which effects are proportional to the causes, nonlinear dynamics describe a system that is sensitive to initial conditions, where very small differences at the outset may lead to vastly different results. Acknowledging this concept forced us to analyze our professional writing program closely, for we couldn't assume simple, one-to-one cause-and-effect relationships in all cases. For example, we might have a variety of workshops to support our teachers; however, one popular, but disgruntled, teacher could have a huge impact on the way a program develops. These were things that we might not be able to necessarily predict, but we could analyze certain actions, certain conditions of the program, to better understand why people were working or thinking in the ways that they did.

Feedback loops can be likened to assessment methods advocated by Eubanks and Abbott (2003) (focus groups); Salvo (2001) (participatory action research); and Johnson (1998) (user-centered design). In fact, much like user-centered design, feedback loops allow all participants in our professional writing program "to take part in a negotiated process of [program] design, development, and use. . . . Users are encouraged and invited to 'have a say,' in other words, and thus they are physically and discursively present in the decision-making processes of [program] development" (Johnson, 1998, p. 32). Johnson is specifically referring to a more informed approach to designing technological systems, but we find his arguments applicable to writing program design (and, hence, our inserting "program" into the previous quotation from Johnson). A feedback loop is a series of actions, each of which builds on the results of prior action and loops back in a circle to affect the original state. The final action either reinforces or changes the direction of the status quo. Healthy systems rely on feedback loops to evolve as a system and coevolve with other systems (the campus and community environment, in our case).

To implement our program, we felt that all stakeholders must assist in its development, to "have a say." This is difficult to achieve in the case of writing programs that rely on rotating staffs of disenfranchised graduate students and adjuncts. Turning to complex adaptive systems theory, we asked ourselves: how could we achieve an autopoietic system, one that has a perpetual, adaptive drive to reproduce itself? How could we integrate the needs and talents of an ever-changing pool of instructors? How could we encourage and sustain professional development in an environment of little support? (Our department does not yet offer, for instance, a for-credit practicum on teaching professional/technical writing.) Praxis as a concept implies a dialectic relationship between practice and theory (Sullivan & Porter, 1997, pp. 25–27). Teachers, like most professionals, are very "praxical." That is, they have a tendency to theorize everyday classroom challenges, in the sense of exercising reflection-in-action, creating models of effective practice, and integrating lore and global/disciplinary theory. We saw this as our opportunity. As we have already noted, technology has already been used in various composition and professional writing settings to facilitate teaching and learning. Technology became a way to focus our efforts on achieving a complex adaptive system, one where the task of designing a useful teaching and learning tool would invest the teachers and students in its design and implementation and encourage them to offer their practical wisdom and other forms of expertise. Moreover, technology offered a practical means to innovative, user-centered assessment practices.

ASSESSING AN INSTRUCTIONAL WEB SITE AS A WAY TO ACHIEVE TECHNOLOGY AS A SITE OF PRAXIS

Because our program goals go beyond simply having a consistent curriculum, that is, because we wanted to use technology to foster our own teachers' professional development and encourage their participation in the development of the curriculum, we sought to develop our own, locally constructed instructional Web site. One way to develop the Web site, we thought, was to "outsource" its development, so we began to discuss our vision for a student- and teacher-centered instructional Web site with publishers at conferences. We realized that there was considerable risk involved with this approach. Namely, we risked losing creative control and intellectual ownership of the project. However, we understood that any royalties could be returned to the program to support the professional development of our graduate students and instructors as well as our ongoing assessment activities.

The conversations we had with publishers were disappointing, and in retrospect, somewhat amusing. We explained our vision to several excited, sometimes befuddled, publisher's representatives, who busily took down notes and who promised convincing demonstrations to follow. Later, we would be sent "mock up" Web sites that were essentially WebCT or Blackboard course-management sites with the names of the main links we suggested added to the standard WebCT interface. This was unsatisfactory. Our own campus was heavily invested in WebCT, so we didn't see this approach as much of a value-added proposition. Moreover, we were cynical about WebCT's potential as a transparent communication medium. In other words, one of our pedagogical goals when using electronic writing technologies is to foster critique of the tool itself, to discuss with students how the technology opens up or constrains discourse. When a campus adopts a particular course-management software platform, its use becomes ubiquitous, comfortable, and invisible.

One series of meetings we had was promising. A representative of Kendall/ Hunt's custom publishing division listened to our objectives and pledged to show us something more than a prefabricated course-management site. At the subsequent demonstration, we were shown an original Web site design, created by one of the company's own designers. The mock up was crude, but was closer in resemblance to Kramer and Bernhardt's Glyph site than anything we had seen before. The publisher also assured us that we could make changes at our discretion, ranging from as small as fixing typos and broken links to as large as adding new content. The ability to make frequent changes was important to us, for given our goals, it was essential to have a dynamic teaching and learning resource.

With input from us and some of the program's teachers, Kendall/Hunt's development team produced a prototype interface that supports our project-based approach to teaching professional writing. As with the Glyph site, completing projects via the Web site "is an exercise first in gathering information and then formulating solutions to the rhetorical problems contained within that information." Students must "move across applications, files, and windows" to "assemble a solution" (Kramer & Bernhardt, 1999, p. 321). The site was initially structured in two layers: (1) a main layer of information that includes a course overview/introduction, links to course principles, links to course projects, a "resources" or handbook section, an index, and instructor contact information; and (2) a second layer for project navigation, including project overview, planning, and templates pages (see Figure 1). The project navigation varies somewhat per project but includes more detailed information about documents students must produce to complete the assignment, samples of student work (with author permissions secured), and evaluation criteria.

In fall 2001 we piloted the instructional Web site in all 20 sections of our program with the intention of conducting formative assessment on its efficacy. We knew that we needed to assess how closely our expectations of the Web site's usability meshed with the students' and instructors' experiences with using the Web site. This programwide implementation was too optimistic and somewhat pushed by the publisher. We wound up spending much of the semester editing content at a pace just in front of our teachers and students. In addition, the publisher failed to implement a planned password-protection system, which resulted in much initial confusion and resentment among the majority of students who purchased registration codes to a site seemingly without a registration system in place. The publisher eventually installed the password system midway through the semester, causing headaches for students who had become accustomed to accessing the site without first passing through a login screen.

At the end of the semester, we conducted an online exit survey of students in all sections and solicited feedback from our instructors. These methods were intended to provide both formative assessment data (feedback to assist in the ongoing development of the instructional Web site) and summative assessment data (feedback on how well the program was achieving its curricular goals in that particular semester). The student survey, which consisted of 50 close-ended questions and 6 open-ended questions, demonstrated that students

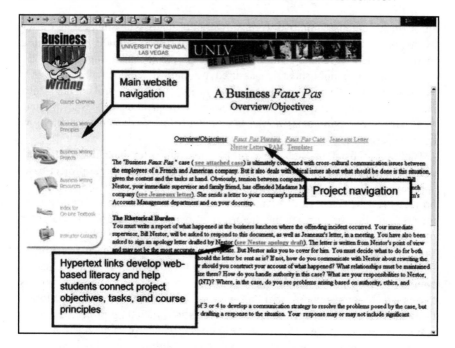

Figure 1. "A Business *Faux Pas*": Sample project from the
Business and Technical Writing WebCOM Web site.

generally thought the computer-networked nature of the course did improve their preparedness for writing in professional settings (survey response rate was 67%, or 255 out of 379 total students). We attributed the publisher's technical glitch with the password system as a factor contributing to nearly 40% of students not liking the online format (see Table 1).

We also found correlations between positive and negative attitudes among students based on instructor. Students for one instructor, who was on probation in our program based on student complaints, overwhelmingly disapproved of the online format and responded negatively to questions about course outcomes. On the other hand, students in our own classes (we also taught in the program), felt strongly in favor of the online format and learning outcomes. This was a phenomenon we had not previously observed in the literature of writing program assessment. This phenomenon, however, was the kind of nonlinear occurrence that we could not necessarily account for in our initial program design, but something we became aware of through our use of complex adaptive systems theory. We believed that this would allow our program to adapt more efficiently, however. By presenting these findings to our instructors in a sensitive way so as to avoid the appearance of using assessment as a tool for surveillance, we made

Table 1. Student Survey Results Comparison

Question	Strongly agree 1	Agree 2	No opinion 3	Disagree 4	Strongly disagree 5	Total Total %
1. I liked the online textbook.						
Spring 2003	41	103	25	44	11	224
	18%	46%	11%	20%	5%	100%
Fall 2002	21	83	27	45	34	210
	10%	40%	13%	21%	16%	100%
Spring 2002	40	114	23	42	22	241
	17%	47%	10%	17%	9%	100%
Fall 2001	15	116	26	49	49	255
	6%	46%	10%	19%	19%	100%
2. I prefer the online textbook to traditional print textbooks.						
Spring 2003	38	59	41	58	28	224
	17%	26%	18%	26%	13%	100%
Fall 2002	20	39	39	48	64	210
	10%	19%	19%	23%	30%	100%
Spring 2002	32	71	60	43	35	241
	13%	29%	25%	18%	15%	100%
Fall 2001	23	61	48	72	51	255
	9%	24%	19%	28%	20%	100%
3. The assignments improved my writing						
Spring 2003	70	106	28	16	4	224
	31%	47%	13%	7%	2%	100%
Fall 2002	40	104	25	31	10	210
	19%	50%	12%	15%	5%	100%
Spring 2002	65	116	39	15	6	241
	27%	48%	16%	6%	2%	100%
Fall 2001	50	122	41	28	14	255
	20%	48%	16%	11%	6%	100%
4. I feel that this professional writing course has prepared me to write effectively in professional settings						
Spring 2003	67	125	24	5	3	224
	30%	56%	11%	2%	1%	100%
Fall 2002	43	111	31	15	10	210
	20%	53%	15%	7%	5%	100%
Spring 2002	68	120	29	12	12	241
	28%	50%	12%	5%	5%	100%
Fall 2001	67	124	34	18	12	255
	26%	47%	13%	7%	5%	100%

Total Response Rates: Spring 2003: 65% (224/342 students); Fall 2002: 55% (210/380 students); Spring 2002: 77% (241/315 students); Fall 2001: 67% (255/379 students).

this trend observed in the data explicit and solicited feedback for better understanding of why it might occur and what instructors needed in the way of professional development to feel more comfortable with the instructional Web site.

In spring 2002, we adhered to our schedule and focused on revising the Web site and conducting an external assessment grounded in usability testing methods. Graduate student teaching assistants who were also teaching professional writing and piloting the Web-based materials helped us refine the curriculum materials. This activity was central to our aims of using technology as a site of praxis and further provided an alternate feedback loop. Graduate students wrote projects (e.g., the one pictured in Figure 1) and other curriculum materials, compiled a detailed index, and researched and annotated links to supplemental Internet resources. We also conducted another online survey of students across all sections. The results of this survey (77% rate, 255 of 379 students) gave us some comparative data from the fall semester and generally demonstrated improvements from the previous semester, in part because of the contributions of the graduate assistants (see Table 1).

Consistent with our commitment to praxis-oriented assessment methods, in summer 2002, we invited an external expert in human-computer interaction, Bill Hart-Davidson (Michigan State), to conduct a usability study of the Web site focused on assessing the following:

1. The usability of the Web site as an online learning resource for students
2. The usability of the Web site as a pedagogical resource for instructors
3. The role of the Web site in furthering the pedagogical goals of our professional writing program

In addition to these assessment goals, the consultant was also asked to provide formative data and recommendations for improving the Web site based on these goals. Accompanied by an assistant, the consultant visited UNLV and conducted usability tests with 10 UNLV student volunteers. The tests involved observing students perform essential tasks related to the Web site and asking them questions before, during, and after the tasks. The consultant also conducted a focus group interview with 5 program instructors to collect their feedback about the Web site as a pedagogical resource. This external evaluation was necessary for a number of reasons, including how a focus on usability encouraged stakeholder participation in the assessment and revision of the Web site. We also felt our instructors would be more comfortable providing criticism to an outside consultant. Furthermore, we could justify paying the instructors for the afternoon they spent with the consultant.

Hart-Davidson submitted a 31-page report, concluding that students found the project layer of the Web site well developed and relatively usable. Several students mentioned the "planning" section of the project pages as being particularly valuable. The students appreciated the multitasking environment fostered by the site, rating the Web site high compared with a traditional textbook for providing all the necessary information to complete a project (4.2 mean on a 5.0 scale). Conversely, students found it difficult to navigate across projects and the primary layer of the Web site. Students had difficulty locating a particular MS Word template and, once found, had difficulty downloading and manipulating it. Hart-Davidson's report offered a number of ways to improve the structure and navigability of the Web site.

As for the instructors, the consultant found that all 5 who participated in the focus group session, which included an instructor-oriented usability evaluation, generally favored the online format. The majority (4 out of the 5) stated they would use our instructional Web site or a comparable one again, even if they were going to teach at another institution. The most common critiques by the instructors included a desire for more content, including background readings and student samples. Instructors also wanted more customizability. They wanted to be able to post their own syllabi and select/show only projects they planned on using, for instance.

As the results from question 1 in Table 1 show, by spring 2003, 64% of the students liked or strongly liked the online delivery of course information. From fall 2002 to spring 2003, the numbers of students not liking the online format dropped from 37% to 25%. These numbers reflect in part the majority of students' inexperience with and ambivalence toward online course formats. While 79% of students in spring 2003 reported having previously taken a class in a computer classroom, 74% reported never having used an online textbook before. One could expect mixed results, then, for a question such as number 2, which asked students if they preferred the online textbook to traditional print textbooks. Of the responses, 43% preferred the online textbook, while 42% did not. Open-ended comments from students indicate that student attitudes toward the textbook are tied to preexisting experience with computers. For instance, students who liked the Web site gave reasons such as "on the computer a lot already," "textbook no longer competing with computer workspace," and "easy to multitask between Word docs and textbook." Students who did not like the Web site gave reasons such as "don't have easy access to computers," "prefer to study away from computer," "don't like reading online" (see Figure 2). These cumulative results, compared with the first semester, helped us modify our earlier assumption that technical glitches were contributing to negative responses from students. Rather, we were able to see a pattern across semesters that had more to do with student comfort with technology in general. However, even as the students adjusted to the Web-based format for course-information delivery, the students still generally felt that (1) their writing skills improved as a result of the course and (2) the course was preparing them for writing in professional settings (see questions 4 and 5 in Table 1). Only 1% of the students felt strongly that they were not prepared to write for work in the spring 2003 survey.

Reasons for liking	Reasons for not liking
On computer a lot already	Didn't have easy access to computers
Glad not to have to carry heavy textbook	Preferred to study away from computer
Easy to multitask between Word docs and textbook	Don't like reading online text
Could access from home, computer labs, library	Have slow dial-up connection and have to pay per minute
Usable organization and navigation	Prefer to annotate while reading
Lots of examples, templates, and links to resources	Printing pages is tedious
	Web site doesn't work on Macs
Textbook no longer competes with computer workspace	Server errors
Costs less than print textbook	

Figure 2. Student reasons for liking/not liking online textbook.

ACHIEVING A "HEALTHY" CULTURE OF ASSESSMENT

Recall from our previous discussion that our assessment activities were influenced by three key concepts drawn from complex adaptive systems theory: autopoiesis, nonlinear dynamics, and feedback loops. Simply put, an autopoetic system reproduces itself. It was our intention from the start to integrate assessment activities into the ongoing development of the professional writing program. As Allen writes, "effective, systematic assessments require more than a one-time offering. They require a commitment to time" (1993, p. 5). Rather than view assessment as something we "have to do" for external audiences on occasion, and rather than have our teachers and students see assessment as additional "work" with no apparent payoff, we sought to incorporate assessment into the culture of our program, to make it a component necessary to the continuance of our curricular system. We believe we have accomplished this through the ongoing development of a local, custom-published online instructional Web site. Because our student exit surveys are online, they can be easily tabulated and presented to various constituencies for analysis (formative feedback) and consumption (summative feedback). Reviewing the results from last semester's student exit surveys has become an anticipated agenda item at our beginning-of-the-semester staff meetings. The surveys help confirm what the teachers experience in isolation and anecdotally report. When viewed as summative evaluations, the surveys help demonstrate progress, which motivates repetition. The trends from the student exit surveys help show the instructors that we are making improvements and that the semester-to-semester bugs and kinks that are a natural part of the interface-development process are worth the inconveniences and apparent inefficiencies. Likewise, and perhaps more importantly, we can present the survey data to incoming students as a way to orient them to the pending learning experience.

One of the key findings from our 3+ years worth of assessment data is that, as we stated earlier, student experiences with the instructional Web site and the course overall vary markedly, depending on instructor. This phenomenon is an example of nonlinear dynamics. We assumed early on not to expect that all teachers would use technology in the same, predictable ways. This suggests that instructor-oriented usability is just as important as student usability and that initial and follow-up training is an essential aspect of moving instruction to the Web. Based in large part on the results of our external assessment conducted by Hart-Davidson, the interface of the Web site was extensively revised in fall 2003 (see Figure 3). The publisher, Kendall/Hunt, incorporated dynamic capabilities

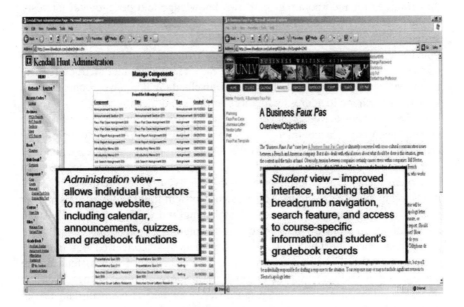

Figure 3. *Business and Technical Writing WebCOM* Web site Version 2.0—
"Administration" and "Student" views.

more in line with our original vision and our own instructors' feedback. This includes an "administration" side that allows for customizable features similar to course-management software like WebCT, including "calendar" and "announcement" sections for posting assignments, with links to parts of the textbook, message board and e-mail listserv capabilities, a testing function, and a gradebook (see Figure 3). The addition of these new tools, however, significantly raised the technology learning curve for instructors. The publisher provided print documentation and onsite training at the beginning of the fall 2003 semester, but the most recent feedback we've collected indicates that these measures are not enough to support some of our instructors who are less comfortable with using technology.

We hoped that through Web-based user surveys and other data-collection methods we could collect rich information about the program, which could be fed back into our curricular system. We feel strongly that the ongoing development of the instructional Web site not only necessitates feedback loops, but that the activity itself creates occasions for dialogue among program constituents that might not otherwise exist. The survey and focus group assessment methods fostered community among our program teachers and encouraged their professional development. Early data indicated that while the main components necessary for our curriculum were in place, our initial product was incomplete in many ways. Our teachers recognized gaps in the site's content and saw opportunities for incorporating their own knowledge and pedagogical interests. Graduate students who contributed to the site and later graduated reported having stronger vitas and more confidence as a result of their participation in the instructional site's development. Current graduate students continue to volunteer to take on new projects related to enhancing the site. Our adjunct instructors, who usually teach about 25% of the total courseload, make fewer contributions, something we hope to address with more rewards and incentives.

Our assessment activities have also affected the relationship between us as authors negotiating a contract with a book publisher. We like to take some credit for getting Kendall/Hunt into the online textbook market, for while they originally sold us with their initial design ideas, we subsequently learned that they were themselves entering uncharted territory in an effort to secure sales rights to our sizeable program. The password system failure of the first semester was only one example of a series of publishing and editing snafus. We spent an inordinate amount of time correcting formatting changes, including typos, introduced when the publisher's Web designers first converted our material, submitted as Word documents, to HTML format. We also had to fend off an attempt to drastically restrict the original promise of unlimited changes, settling for a system of bi-weekly changes. Bolstered by our assessment-data driven arguments, we gradually persuaded the publisher to expand its narrow notions of online publishing and its technological capabilities. The publisher originally could not provide us with the dynamic, customizable environment we desired, but the latest version moves significantly closer. The latest version of the instructional Web site even permits differing levels of administrative access, allowing us to make editorial changes to the Web site, such as fixing broken links, thus eliminating the mediation of the publisher's Web designer at later stages.

CONCLUSION

We hope that through this discussion, other administrators of professional writing programs will be encouraged to (1) see the connections between complex adaptive systems theory and user-centered design and how these concepts can positively contribute to understanding assessment as integral to the systematic design of a professional writing program; and (2) see how technology can facilitate feedback loops that contribute to the healthy evolution of a system of teachers and students. Our use of praxis-oriented assessment methods focused on program development—as opposed to student or teacher evaluation—and brought the stakeholders more firmly into the development process at all three stages. Our use of multiple methods (survey and focus groups, usability testing, and external evaluation) created rich feedback loops, which enhanced our vision for a coherent and complex writing program.

We are satisfied overall with how far we've come since fall 2001 and our publisher's commitment to developing our instructional Web site according to our needs. As we discussed, using technology to improve instruction has been achieved to some extent. We hoped that the online instructional space would encourage teachers to share and extend their collective knowledge. We are pleased at the level of contribution from our graduate students, but wish to extend this to the handful of adjuncts who teach in our program. Even with the new dynamic administrative capabilities of the latest version of the Web site, its structure still impedes instructors from sharing innovative ideas and adopting assignment variations. Most of this lore is shared via a program listserv. Recently, however, we have added an "instructor resources" section, which can be hidden from the student-user view and that only instructors can access. This physical space on the Web site has made it possible to share the teaching guides and other information that we formerly distributed in print form. With the addition of this space on the Web site, we have seen an increase in the pedagogical sharing among program teachers.

One limitation of the current dynamic version of the Web site remains the different levels of access based on "user group"—student, instructor, administrator (us), and publisher/developer. While instructors can customize such components as exams and announcements, they cannot yet customize the text of projects and other content. Instructors, in other words, do not yet have the capabilities of literally rewriting certain parts of the text to suit their teaching preferences. We are still a ways from achieving a truly dynamic instructional Web site. The publisher has yet to resolve the technical problems with producing such a customizable text available to multiple instructors and hundreds of students, and we ourselves have yet to fully resolve the pedagogical dilemma of academic freedom within the context of curricular consistency.

Notwithstanding the need to resolve these long-term obstacles, we see the program as fast becoming a self-producing, complex adaptive system, independent from individual influence and organized so that assessment procedures will continue to sustain and shape the program in significant ways. The program has established a firm niche in the larger institutional structure, and our assessment activities will act as catalyst for the program's own growth and development as an integral part of the university.

REFERENCES

Allen, J. (1993, September). Assessment methods for business communication: Tests, portfolios, and surveys. *Bulletin of the Association for Business Communication, 57*(3), 1–5.

Cook, K. C. (2003). How much is enough? The assessment of student work in technical communication courses. *Technical Communication Quarterly, 12*, 47–66.

Coppola, N. (1999). Setting the discourse community: Tasks and assessment for the new technical communication service course. *Technical Communication Quarterly, 8*, 249–268.

Dias, P., Freedman, A., Medway, P., & Paré, A. (1999). *Worlds apart: Acting and writing in academic and workplace contexts*. Mahwah, NJ: Lawrence Erlbaum Associates.

Eubanks, P., & Abbott, C. (2003). Using focus groups to supplement the assessment of technical communication texts, programs, and courses. *Technical Communication Quarterly, 12*, 25–46.

Hundleby, M., Hovde, M. R., & Allen, J. (2003). Special issue on assessment in technical communication. *Technical Communication Quarterly, 12*.

Jablonski, J. (2004). *Business and technical writing WebCOM*. Dubuque, IA: Kendall/ Hunt.

Johnson, R. (1998). *User-centered technology: A rhetorical theory for computers and other mundane artifacts*. Albany, NY: State University of New York Press.

Kemp, F. Instruction manual for TOPIC. Retrieved September 30, 2003, from http://ttopic.english.ttu.edu/manual/manualreadall.asp?typeof=manual

Kramer, R., & Bernhardt, S. A. (1999). Moving instruction to the Web: Writing as multitasking. *Technical Communication Quarterly, 8*, 319–336.

Marion, R. (1999). *The edge of organization: Chaos and complexity theories of formal social systems*. London: Sage.

Mingers, J. (1995). *Self-producing systems: Implications and applications of autopoiesis*. New York: Plenum Press.

Nagelhout, E. (1999). Pre-professional practices in the technical writing classroom: Promoting multiple literacies through research. *Technical Communication Quarterly, 8*, 285–299.

Porter, J. E., Sullivan, P., & Johnson-Eilola, J. (2001). *Professional writing online*. Boston, MA: Allyn and Bacon.

Rasch, W., & Wolfe, C. (Eds.). (2000). *Observing complexity: Systems theory and post-modernity*. Minneapolis, MN: University of Minnesota Press.

Rose, S., & Weiser, I. (Eds.). (1999). *The writing program administrator as researcher: Inquiry in action and reflection*. Portsmouth, NH: Heinemann.

Rose, S., & Weiser, I. (Eds.). (2002). *The writing program administrator as theorist: Making knowledge work*. Portsmouth, NH: Heinemann.

Salvo, M. (2001). Ethics of engagement: User-centered design and rhetorical methodology. *Technical Communication Quarterly, 10,* 273–290.

Sullivan, P., & Porter, J. E. (1997). *Opening spaces: Writing technologies and critical research practices*. Greenwich, CT: Ablex.

CHAPTER 12

Reconsidering the Idea of a Writing Program

William Hart-Davidson
Michigan State University

There are two exciting and potentially controversial perspectives on assessment taken by Jablonski and Nagelhout that should be the subject of conversation, in my view, for all professional writing program administrators and faculty. I paraphrase their main points, as follows:

1. The program, itself, ought to be assessed as a unit; such assessment may involve measuring student performance and outcomes, but should not be limited to these alone.
2. A site, and perhaps *the* site to conduct programmatic assessment, is the technology that makes the program what it is.

I suspect that while the first of these statements is something most Technical and Professional Communication faculty and administrators would readily agree with, it is probably not carried out very often, except as part of other comprehensive assessments such as accreditation reviews. Making a commitment to program assessment of the type that Jablonski and Nagelhout describe—not "something we 'have to do' for external audiences on occasion" but "a component necessary to the continuance of our curricular system" (p. 182)—is perhaps easy to do, but tough to keep up on an ongoing basis. And one reason why it is tough to keep is the same reason Jablonski and Nagelhout give as a rationale for making such assessment mandatory: "a writing program is a complex curricular system of students and teachers" (p. 172). How and where do you go to have access to this system and what goes on within it, particularly when it is distributed across many sections? And, in the case of UNLV's program of commuting and distance students, what are the effects across many places and times of day?

Jablonski and Naglehout's answer to this last question is to look at the online system. In effect, they suggest that "the system" in the abstract and "the system" created as an online course-management and delivery resource for teachers and students can be the same thing for the purposes of programwide assessment. To oversimplify: you are your Web site. Or, perhaps a bit more accurately, your program consists of the technological resources—and the interactions and information that flow across them—that you make available to stakeholders.

I suspect that this idea is a bit tougher to accept than the proposal for consistent and integrated program assessment. As humanists, Technical and Professional Communication faculty and administrators tend to be a bit wary of reducing a program to its technological components and extremely careful to portray a program in terms of the people who compose it. What Jablonski and Nagelhout begin to show us here, however, is that a careful look at the technological system can be a careful and responsible look at the people in a program as well.

For those who remain skeptical, though, about both the value and ethical fitness of examining technology as a way to understand what is working and what needs improvement in a Technical and Professional Communication program, I'd like to offer another rationale in this response. The rationale I will offer in the form of a broadened mandate for writing programs builds on Jablonski and Nagelhout's arguments that our programs are "the property of all stakeholders" (p. 171), and that as complex, dynamic systems, they have motivations to exist beyond their responsibilities to external groups (p. 173).

BROADENING THE MANDATE FOR WRITING PROGRAMS

I believe that writing programs have a responsibility to grow the capacity of their stakeholder communities, not merely individuals within them, to produce good writing and good writers. As a fundamentally social activity, good writing does not happen within individuals alone, and the capacity for good writing performance does not develop in sustainable ways in bursts of time as short as academic semesters. We can never say that a student is "finished" learning to write, no more than we can say a piece of writing is, for all time and all situations, good beyond improvement.

While much of our theoretical, curricular, and pedagogical efforts are focused on the way individuals—our students—develop as writers, we should, as Jablonski and Nagelhout imply in their discussion of programmatic assessment, also be focused on how our programs contribute to the social collectives that they are situated within. Doing so, I would argue, significantly broadens the scope of responsibility that writing programs assume in their departments, on their campuses, and in regional, national, and global settings as well. If we take seriously the idea that our programs should make communities better at writing, then we can begin to see how offering a technological infrastructure for supporting writing performance and learning move to the center of a program's mission. In fact, we might imagine a writing program in this vein that chooses not to offer "courses" or explicit instruction at all, but instead focuses on enabling the many-to-many interaction needed for communities to foster development in writing ability. This sort of program might offer technological infrastructure as its primary, if not its only service.

If this idea seems far-fetched, consider that "the idea of a writing center" was once considered a radical one (North, 1984). Writing centers are now quite common companions to traditional writing programs, offering a kind of support distinct from their traditional counterparts by varying the mode of interaction that they are built upon. Instead of the one-to-many model of classroom-based writing instruction, they offer one-to-one interaction. They zero in on the individual writer dealing with the particulars of a rhetorical situation. And they provide valuable, timely, and individualized assistance that supports but does not duplicate the assistance that students get in classroom settings.

My proposal here is similar to the idea of a writing center in that I am suggesting we consider another variation to the interaction model for writing programs: many-to-many. I believe that for a variety of reasons related to the transformation of the U.S. economy to an information-centered enterprise, academia and society are ready to consider writing programs that adopt this model, even though it suggests very different goals for writing programs than those we have seen develop in the last century. I offer the following list of trajectories of program development to consider:

- A move away from the individual and toward the social group (community of practice) as the primary locus for rhetorical inquiry, measuring rhetorical effectiveness, and fostering rhetorical change
- A move away from an emphasis on rhetoric as enacted in discursive "moves" made by individuals and concretized in the utterances of those individuals and toward an emphasis on rhetoric as patterns of activity enacted across specific discursive events by groups or communities of practice
- A move away from a view of texts as either snapshots of past action or indicators, however partial, of action in the present and toward a view of texts as locations for ongoing and future interaction among members of a social group
- A move away from the study of rhetoric(s) as windows on more-or-less fixed social orders (e.g., "the rhetoric of science") and toward the study of

rhetoric(s) as dynamic collections of discursive "resources" available to members of a community in a given social-historical moment

Each tries to characterize what a larger move toward stakeholder-centered writing programs might entail. They are meant not to leave the individual student writer behind, but to reconsider the place of the individual writer as the sole focus of a writing program.

I do not have the space here to flesh out the trends in research, curriculum, pedagogy, economic and technological development that might lend support to these trajectories as actual paths that writing programs are or will be taking. But Jablonski and Nagelhout talk about a few such trends that are worth noting, including the move to offer more sections via distance and the related attempts to provide more and more students with options to access higher education. Perhaps most significant, though, is the idea that students need to develop multiliteracies for information-centered careers and lives as citizens in a digital world. This means that learning to write (and here, I use "write" as shorthand for the range of performance skills that include the ability to invent, arrange, style, store and retrieve, and present compelling texts and images) is a lifelong prospect. Learning to do these things well and to continually adapt to new media forms and situations for doing them is not something that can be done in 10 weeks, 10 months, or even 10 years. Given that this is the milieu our students occupy, shouldn't our programs be there for them wherever and whenever they need to learn? Within a traditional programmatic structure, support for lifelong learning in writing is simply not feasible. But it is entirely feasible in the digital environment.

So the question I'd like to pose here is this: If we have the capacity to support lifelong learning in writing, and if both our theoretical foundations and research results suggest that such support is needed, do we have the resolve to act, as programs, to provide such support? If we do, we will need to understand what Nagelhout and Jablonski have learned in their efforts to build their program at UNLV: building the technological infrastructure is at the center of the program's goals.

OUR NEW MISSION:
SUPPORTING THE SOCIAL ACTS OF WRITING

I want to acknowledge that I have not limited my discussion of a broadened mandate to T&PC programs, alone. I have spoken, rather, of "writing programs" more generally. I chose to do this because the conditions under which first-year writing programs and technical and professional writing programs grew up as distinct entities are severely undermined by the same trends mentioned above. It is becoming clear that being a citizen in the digital age requires the range of multiliteracies that we may once have considered specialized knowledge for workers in high-tech industrial sectors. I do not see much of a future for writing programs that maintain divisions based on what genres of writing are taught or what media are involved. Levels of emersion and specialization may be maintained for, say, students majoring in professional writing vs. those majoring in mechanical engineering, but the need for continual learning across a broad spectrum of genres and media will remain great for both.

As my good friend and colleague Karen LeFevre taught us, invention is a social act (1987). I am hereby suggesting that arrangement, style, memory, and delivery are, too. And if we think carefully about how the actual work of the rhetorical canons is carried out today—how texts are stored and retrieved in computer memory, for example, or how students in distance sections share drafts for review and then turn in revised versions without ever meeting their teacher or peers face-to-face—we can see how central a role the technological infrastructure that supports our writing programs are already. We are, indeed, our Web sites, our e-mail lists, our server spaces, and file-sharing protocols. These are the spaces but also the means by which writing is carried out and learned. These are also the spaces and means by which we have built and will sustain an information economy. The place of writing programs in the information economy might therefore be understood as providing "support," understood in both the familiar and an unfamiliar sense. A support role is familiar, and not unproblematic, if the circumstances for learning to write are understood as imparting "basic skills" that can be mastered and, presumably, left behind as students move on to more advanced learning. It is unfamiliar, on the other hand, if "support" means that writing programs might provide the infrastructure that allows communities to develop and deliver anytime, anywhere assistance for writers confronting a rhetorical situation. It's the latter flavor of "support" that Nagelhout and Jablonski's view of program assessment and iterative development point us toward. And it is this view, also, that promises to ensure the continuing relevance of writing programs in the information economy.

REFERENCES

Lefevre, K. B. (1987). *Invention as a social act.* Carbondale, IL: Southern Illinois University Press.

North, S. (1984). The idea of a writing center. *College English, 46*(5), 433–446.

Assessing Intercultural/
International Projects

CHAPTER 13

Assessment in an Intercultural Virtual Team Project: Building a Shared Learning Culture

Doreen Starke-Meyerring, McGill University
Deborah C. Andrews, University of Delaware

This volume addresses two central questions: *how* do we assess the work of our students and *why* do we use assessment as we do. These questions take on new meaning when students and instructors work in a distributed, shared virtual learning environment. And that new meaning becomes even more complicated when a classroom partnership extends across institutional and national boundaries. But such partnered learning projects are becoming increasingly common as technical communication faculty recognize that communicating well in culturally and technologically complex situations is a critical success factor for global work and citizenship (Herrington, 2004; Herrington & Tretyakov, 2005; Sapp, 2004; Starke-Meyerring, Duin, & Palvetzian, 2007; Zhu et al., 2005). Having emerged only recently, these partnered learning environments require new thinking about teaching practices and thus new thinking about how we assess the effectiveness of those practices. In particular, these new learning environments raise questions about how best to assess the benefits and costs of various supporting technologies, to negotiate differing assessment practices across institutions located in different national education systems, and to measure gains in cultural understanding and teamwork skills—achievements that are difficult to measure even in one classroom.

In this chapter, we address these questions by analyzing the role of assessment in a semester-long intercultural virtual team project we developed to connect our two courses in management communication, one in Canada and one in the

United States. Designing the course engaged us in extended discussions about assignments, logistics, and technology use (Starke-Meyerring & Andrews, 2006). We also developed an assessment strategy that was a key element in the partnership's success. We aimed for a process that was dynamic, formative, and interactive to facilitate ongoing student efforts at building a shared learning culture. In this chapter, we describe our assessment strategy. We first provide a brief overview of the project—its intercultural and virtual aspects that converge in specific challenges for intercultural teamwork. We then show how we aligned our assessment with the learning objectives of the partnership and illustrate the specific assessment approach we used to help students achieve those objectives as well as our decisions and the logistics involved in negotiating our local assessment practices to build a shared approach. Finally, we discuss the outcomes of the project for students as well as for faculty and conclude with recommendations for assessment in future partnerships.

OVERVIEW OF THE INTERCULTURAL VIRTUAL TEAM PROJECT

The partnership project aimed to help students learn and communicate in an intercultural and technologically mediated environment. Here we briefly describe the project we designed for this purpose: the courses, institutions, and students involved; the tasks and structures we put in place; and the project's intercultural and virtual aspects.

Courses, Institutions, Students

The project connected two management communication courses—one at McGill University, Montreal and one at the University of Delaware—that help students shift from communicating in an academic setting to communicating in a business environment. The McGill course, "Communication in Management II," is an upper-level undergraduate course designed specifically for students from McGill's Faculty of Management, who major in such disciplines as finance, accounting, marketing, or international business. The course has been offered for a number of years by the Centre for the Study and Teaching of Writing, which is housed in the Education Faculty. The University of Delaware course, "Written Communications in Business," is similar in purpose but is designed for students across disciplines. Accordingly, students in this course came from a variety of majors, including nursing, agriculture, marketing, and the visual arts. The course is taught in the University's English Department.

Like the courses, the two universities are similar, but do exhibit differences. Both are research universities of roughly the same size. However, as a land-grant state university in a small town, the University of Delaware has a very different institutional culture from that at McGill, an international university in a large and ethnically diverse city. In addition, while the language of instruction at

McGill is English, the University is located in a French-speaking province and in a major French-speaking center of international commerce and culture. All the students in the class were at least bilingual, many spoke French as their first language, some spoke up to seven languages, a number had lived in a variety of different countries, and several were international students with roots beyond Canada (including the U.S.). Only one international student—from China—was enrolled in the Delaware class.

Tasks and Structure

Adapting a model developed by Bertha Du-Babcock at the City University of Hong Kong and Iris Varner at Illinois State University, we designated seven student teams of six to seven members and asked them to examine business communication practices at enterprises with operations in both Montreal and Delaware (or environs). We assigned students to the teams randomly, with a more or less equal number from each institution on each team. The teams chose their own leaders and vice team leaders, but we stipulated that four teams would have leaders from McGill (with vice team leaders from Delaware) and three from Delaware (with vice team leaders from McGill). We provided a list of suggested industries and enterprises to examine, but the teams were free to choose their own subjects. Six of the seven teams chose U.S.-based retail brands: Ben & Jerry's, Courtyard by Marriott, The Gap, Krispy Kreme Doughnuts, Starbucks, and Wal-Mart; the other chose a Swedish company, Ikea.

The partnership that made this project possible had grown out of a personal conversation at a professional meeting that led to a visit by Andrews to give a workshop at McGill. E-mails followed, then another conversation in Montreal, and then we started designing the joint project. In doing so, over the months before the class and during the class itself, we conducted a related virtual project of our own as we created a shared *instructional* culture that would sustain a shared learning culture for the students.

After negotiating and stating a set of learning outcomes, we designed specific assignments and wrote extensive specifications for each assignment, which we then posted on a Web site shared by both classes. These assignments simulated the traditional genres that help manage complex collaborative projects in the workplace: statement of team policies and procedures, proposal, progress report, oral report (given separately by each team in its own local classroom), final report, and reflective memo. In addition, we asked students to set up a team blog (using blogger.com) and to post regularly to the blog.

The Intercultural Aspect of the Project

Our courses and their locations in different national, institutional, and regional settings provided us with a rich intercultural context in a number of ways. While Canada and the United States share an extensive border and (on the surface) may appear similar in many ways, both are, of course, sovereign nations, each with its own set of political, legal, and cultural practices that reflect diverse values, assumptions, and traditions. A quick survey of headlines in either country's media will readily reveal such divergent values, which may render practices that are considered legal and ethical in one country problematic in the other. To provide only one example, in response to the U.S. Patriot Act, which requires U.S. companies to disclose personal data about individuals at the government's request, recent Canadian laws prevent Canadian universities from storing student data on servers located in the United States (such as those of plagiarism-detection services or e-learning providers) and professors from carrying laptops with student data across the border (Keller, 2007). This example may suffice to illustrate the ways in which values, priorities, and assumptions can vary across national contexts.

However, the situatedness of our classrooms in different national contexts was only one aspect that contributed to the richness and complexity of our shared learning environment. In addition, students, especially at McGill as one of the most international universities in North America, represented a number of national, ethnic, and linguistic backgrounds and experiences, while Delaware students represented wide-ranging professional diversity. This professional diversity, for example, was particularly important for management students, who need to learn how to communicate about business issues with a variety of professionals, yet in their degree programs are often exposed mostly to other management students with similar professional backgrounds. Moreover, the Delaware students contributed professional perspectives from multiple disciplines. At the same time, the McGill students, who all majored in various aspects of business studies, contributed specific business perspectives.

From the beginning, we assumed that identities are complex and extend far beyond particular national affiliations or identifications, and that the interaction between cultural identities and communication is equally complex, rarely allowing for simple or easily separable one-on-one matches between cultural context and communication choice. With Scott, Longo, and Wills (2006), we resist views of culture as the simple homogeneous properties of groups, but understand culture rather as consisting of regularized, habitual daily practices and values that are constantly shaped and contested in complex ways by individuals and groups situated in any number of contexts. Such a view of culture, for example, does not equate nationality with culture, but assumes that individuals may or may not identify with or may actively contest dominant national practices, policies, or decisions. Learning about the ways in which national contexts interact with business communication practices was therefore only one of a number of objectives, since the main goal was helping students learn how to negotiate diverse practices and values regardless of their origin, and to navigate, communicate, and thrive in a culturally rich and distributed environment.

This intercultural aspect of the project required the building of a shared culture of intercultural learning—one in which students could learn to appreciate the complex identities of individuals, to put inquiry before rash judgments or attributions of motivations or intentions to others, and to learn with and from people representing diverse cultural identities and contexts. Given our physical locations, this culture of learning had to be built online.

The Virtual Aspect of the Project

Digital technologies have led to increased opportunities for collaboration across borders both in academe and in other professional workplaces. Organizations have found significant advantages in such virtual collaboration, especially the ability to tap diverse skills and local experience from around the globe while avoiding the costs of travel and compensating for differences in time zones. Accordingly, we wanted to take advantage of the team project to help students develop the digital literacies they need in order to thrive in such technologically mediated environments. These include not only functional literacies such as learning how to use the technologies effectively but also critical and rhetorical literacies (Selber, 2004) such as critically assessing how technologies influence, enable, or constrain communication.

For this purpose, we asked the teams to use and critically assess a variety of technologies for communication in their virtual teams. We had anticipated that the major form of team communication would be Web logs (blogs) because they are free, accommodate links to sites for background information and research entries, and allow for the posting of graphics and other visuals. We also introduced the teams to a Web-based workspace for collaborative writing called ACollab (Accessible Collaboration Environment), an open-source software from the Adaptive Technology Resource Center at the University of Toronto. In addition, the teams used e-mail and instant messaging for more informal communication among themselves and with us. McGill students used a designated WebCT space, although this was not available to Delaware students and thus was not as significant in the project as in their other work in the course.

Both classes met in computer labs. The Delaware lab has rectangular tables, each holding 3 desktop computers, arranged in four parallel rows divided by a central aisle, for a total of 24 computers. Because teams had four members, one always had to sit without a computer or move to a different row to use one. Communication across the team was difficult because there was little room to move within the row, and they couldn't face each other and the screens. The McGill students worked in a wireless site with tables that could be assembled so that team members, each with a laptop, could face each other as they worked. The table also accommodated papers and other materials. It was easier for McGill-side teams to jointly look at one screen as well as to work independently. Photographs of the teams exchanged during the early days of the semester revealed these differences, somewhat accidentally, and caused some envy among Delaware students.

THE CHALLENGES OF VIRTUAL INTERCULTURAL TEAMWORK

To develop assessment strategies that would help us and the students build a shared culture of intercultural learning, we had to consider some of the challenges in virtual teamwork. Cramton (2002) points out five interconnected vulnerabilities that we were alert to in the project. First, team members may fail to communicate and remember contextual information about others at remote locations, including features of the equipment they use, competing responsibilities and pressures, as well as local holidays and customs. Second, the team may distribute information unevenly, with some members not on appropriate distribution lists or not capable of receiving messages as easily or swiftly as others. Third, team members may not see the same information as *salient* if, for example, subject lines on e-mail messages are not changed when new information is introduced into a threaded discussion or a problem or new issue is not properly highlighted. Fourth, team members may differ in their relative speed in taking on tasks and responding to messages because of their own work styles, their differing sense of the priority of the project, their differing access to the technology, and the stability of the technology at their site. Fifth, and a related point, team members may be uncertain about the meaning of silence. Silence may be a message, and if so, the message may be interpreted very differently from what the sender intended. As we discuss below, the students in our classes struggled with all of these challenges.

From an intercultural perspective, however, perhaps the greatest challenge consists of overcoming the formation of locally based subgroups, which naturally have more opportunities to communicate, including communicating face to face within and beyond the local classroom. Cramton and Hinds (2004) argue that correlated characteristics among only certain members (e.g., the shared characteristic of a location) generate faultlines within virtual teams, which can be activated by various events, including task difficulties, policies, or external national or international events. Once activated, they can easily lead to antagonistic behavior, in which subgroups end up pitted against each other, resorting to ethnocentrism, which ultimately hinders intercultural learning. Based on a sense of superiority over the other group, ethnocentric subgroups are less likely to cooperate or to share information with the perceived "other" group (Cohen & Bailey, 1997), leading to communication problems and ultimately failure (Earley & Mosakowski, 2000).

According to Cramton and Hinds (2004), two risk factors increase the likelihood of such faultlines developing and erupting: the geographic distribution of the virtual team across only a few locations (especially only two) and an equal number of teammates in each location, leading to more opportunities for sub-groups to form around the shared location. With roughly an equal number of students in each location, the virtual teams in our project were thus at risk of developing such faultlines.

To overcome these risk factors, team members had to build a shared culture. Communication over the Internet helped but was not enough. Cramton and Hinds (2004) suggest a number of conditions that foster the creation of a shared culture:

- The local groups that constitute the virtual team must have equal status.
- The virtual team must pursue common goals.
- The virtual team must be structured in such a way that teammates are interdependent across locations for the successful completion of the project.
- The teammates must have institutional support for intergroup contact.
- The virtual teams must strive toward inclusive communication, using practices and technologies that automatically facilitate inclusion, keeping all teammates in all locations informed at all times.
- Virtual teams must pay particular attention to sharing contextual information (e.g., about local customs, holidays, or work practices).

Building such a shared culture across locations and across the various regularized and habitual patterns participants would bring to the team, then, is a challenging task with significant risks, and as such formed the main learning goal of the partnership project.

ALIGNING ASSESSMENT WITH THE LEARNING OBJECTIVES OF THE PARTNERSHIP PROJECT

As with any assessment decision in any context, the guiding question is what is to be achieved and hence to be assessed (Ramsden, 1992; Shepard, 2000; Walvoord & Johnson Anderson, 1998). Given the main goal of the partnership project—communicating in a culturally and technologically rich environment—we articulated the following three sets of specific objectives:

1. Develop proficiency in written and oral professional communication genres.
   - Become familiar with the genres and conventions of professional communication and know how and when to adjust them for changing circumstances and technologies.
   - Use various common workplace genres, such as memos, proposals, job application materials, and reports to manage teams and projects.
   - Prepare and deliver effective team oral presentations.

2. Master communication processes and strategies.
    *   Analyze the audience, purpose, goal, outcomes, and media of a communication situation and develop a communication strategy that is appropriate to this situation.
    *   Design and organize documents for usability and readability.
    *   Select, design, and place visuals strategically to convey dense and difficult material
    *   Use appropriate technology for designing and presenting information and choosing the right media for the situation.
    *   Edit for precision, clarity, conciseness, and accuracy.

3. Communicate successfully in intercultural virtual teams.
    *   Apply the principles of effective communication in an increasingly global, technologically mediated, and complex business environment.
    *   Use teamwork to solve communication problems.
    *   Use and critically assess the use of digital communication technologies in managing team communication.
    *   Recognize how various cultural contexts impact business communication practices.

With these objectives in mind, each assignment was then designed to address one or more of these learning outcomes. The team memos and the blog assignment, for example, also specifically addressed virtual work. The policy memo in particular aimed to counteract early on some of the vulnerabilities cited by Cramton (2002) through the development of an explicit team charter. Students negotiated the content and writing of this memo, whose audience was the entire team as well as us. They assessed team expertise at each site, talked about leadership and decision making, reviewed the tasks that needed to be assigned and guidelines for assigning them fairly, established times for each team member to check in with teammates and the technology for checking in, reiterated the deadlines and writing approaches for future assignments, stated team values and norms, and noted actions to be taken if team members did not adhere to the guidelines. With this memo in place, we encouraged the teams to monitor themselves as they worked and to appeal to this statement if things went wrong. The team status memo and the reflective memo were designed to help teams gauge their success at building a shared team culture and to articulate lessons learned.

ASSESSING INTERCULTURAL VIRTUAL
TEAMWORK

Given these learning objectives and the newness of the project both for our students and for ourselves, we took a constructivist approach directed at building a "learning culture where students and teachers . . . have a shared expectation that finding out what makes sense and what doesn't is a joint and worthwhile project, essential to taking the next step in learning" (Shepard, 2000, p. 10). We thus used assessment to help students learn, to report on student progress, and to make decisions about teaching (Ramsden, 1992).

For this purpose, we heavily emphasized formative assessment, in Shepard's (2000) terms, *dynamic assessment,* "a type of interactive assessment, which allows teachers to provide assistance as part of assessment" (p. 10). As Shepard points out, dynamic assessment directly supports learning: "it creates perfectly targeted occasions to teach and provides the means to scaffold next steps" (p. 10). Throughout the course, we assessed students both informally and formally as they built team identity across cultures; developed mutual knowledge of their chosen enterprise; and drafted, revised, and edited their various memos and reports. Given our close work with students and our ongoing dynamic assessment while they wrote in the labs, we had originally intended to give final grades at the original due dates for each milestone document. But a few weeks into the course we realized that our comments in the labs were not enough to ensure quality in the first submitted draft, and that our emphasis on dynamic, formative assessment had to be strengthened beyond our work in the labs. None of the students had ever worked on an intercultural virtual team. In addition, they had to negotiate a number of different preferences, practices, and interpretations of assignments. Most students were uncomfortable with the ambiguity and confusion such negotiation entails. They needed our written review of a whole draft to then work as a team toward a better second draft. We thus adjusted deadlines accordingly and graded only the second submission.

Our comments addressed their research, the structure and expression of their communication product, and most important, their teamwork. We tried to help them develop next steps, reflect collaboratively on their work as they reviewed our comments, and monitor their own dynamics. For the longer report at the end, we provided a summary set of issues they needed to address as well as several models of segments they could adapt to express their own information. In reviewing work from teams that had become well integrated, we articulated and reinforced the signs of that integration. For example, we noted team logos and names that suggested the forming of a shared team identity. One successful team early on called themselves the "McDels" and created a logo for their team—a red hen, combining the mascots of the two schools (the Redmen of McGill and the Blue Hens of Delaware). The Krispy Kreme team kept a name for each side that suggested collaboration: the "Delkrispys" and the "McKremes." Conversely, for teams who seemed stuck at the faultlines of their geographic locations and whose language indicated that they considered themselves two separate teams ("the McGill team" and "the Delaware team"), we provided comments and questions to help them reflect on their team development across locations. Overall, our comments were heavily directed at motivating and encouraging the development of a shared team culture.

Such dynamic formative assessment provided excellent opportunities for teachable moments as students wrote and chatted in the labs, especially as they tried to build good rapport with their real-life audiences at the other site, whose text appeared before them on the screen, awaiting responses. The teams had to negotiate disagreements with choices made in their drafts, suggest alternatives, and in general, talk about what they were writing. One team, for example, became frustrated when the only revising their counterparts undertook was the deletion of a few words. That provided a good moment for a discussion of the context in which their counterparts were working and on strategies the team might use to make explicit what team members could reasonably expect from each other as they reviewed and enhanced their mutual text.

The students themselves were part of the assessment process. For example, in their progress report, they reflected on their ability to build a team culture, to develop structures for their collaboration, and to set and adhere to a schedule. At the end, to emphasize their ability to build a shared virtual learning culture, they wrote a reflection memo focusing on their team dynamics and technology use and capturing the knowledge they developed as a team about communicating and learning in a distributed environment.

NEGOTIATING SHARED ASSESSMENT PRACTICES: LOGISTICS AND BEYOND

To help model a shared culture in the students' work, we decided to respond to that work with collaborative comments on our side as well. We developed the following procedure. One of us took the lead on the assignment (we alternated across the semester), receiving documents as electronic files from each team, making the first comments using MSWord's "Track Changes" function, then shipping the files to the other instructor. She then read the document and the comments and made further comments (usually in another color) and returned the files back to the lead instructor for the assignment, who read those comments, made any additional comments, and sent the annotated files to the student leader for distribution to the team. The process was less cumbersome than its description may suggest and worked well. We usually had comments back to the teams within a few days, often over a weekend. In addition, throughout the course, we checked in with each other at least once a day, usually before and after class sessions, to compare notes on what had worked, or not worked, that day. While the students didn't see our correspondence, it was very much in line with what we expected them to do as they negotiated their work virtually and kept us aware of the potential struggles of such work.

In developing our comments and trying to keep the team members equal at both sites, we also had to negotiate between the assessment cultures of each institution. At McGill, students expected to do much of their work for the project in class; Delaware students were more prepared to work outside of class on team projects. Both classes used the same textbook for the course—*Management Communication*—a brief guide underpinned by an international approach to the management communication process. Concepts and guidelines in the text gave us a framework for our comments. Scenarios helped students apply more broadly the lessons they were learning empirically in the class.

To promote team cohesion and reinforce the equal status and treatment of team members, we assessed all the milestone assignments on a team basis, one grade per team. These assignments made up 50% of the course grade, a high percentage intended to ensure that students would become invested in the team's success. To correct for any unwarranted leveling possibly resulting from this approach and thus to recognize different effort and achievement among team members, we assessed the blog entries individually (15% of the grade). The rest of the grade was derived from an overall professionalism grade (active participation, leadership, collaboration) and an employment communication module with which both of the courses started while the students were setting up their communication technologies and beginning to get to know each other.

In addition to working out the procedures for grading, we also needed to work out the grading system. Each of us has a somewhat different approach to grading and works in a different institutional assessment culture. To arrive at a shared system, we adopted Andrews's way of structuring comments and Starke-Meyerring's point system (Appendix A), which is similar to the 4.00 GPA system. The numbered system then allowed us to convert the shared assessment into the different percentages for grades specified by the grading policy at each institution.

ASSESSING THE PROJECT

Given our constructivist approach, we were looking for outcomes both in student learning and in our own teaching and development as faculty. This section provides our provisional assessment of the project as a whole.

Outcomes for the Students

Every team achieved the stated learning outcomes, though to varying degrees. Despite the challenges, all teams completed all assignments and made progress toward the three major outcomes as indicated by their responses to a pre- and post-project survey (Appendix B) and our comments and grades on their assignments.

DEVELOPMENT OF GENRE KNOWLEDGE IN PROFESSIONAL COMMUNICATION

The survey results indicate that students felt they had considerably increased their proficiency in understanding and applying professional communication genres (Figure 1).

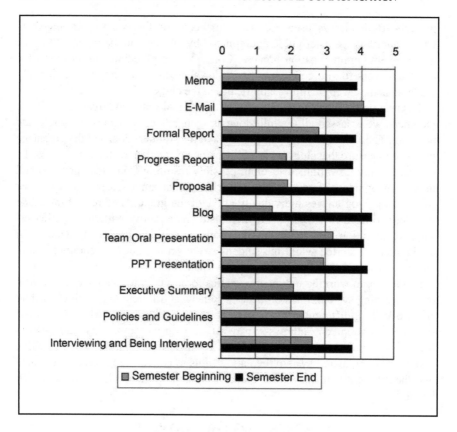

Figure 1. Students' self-rated genre knowledge.

Students brought to the project a variety of perspectives and expectations about business research and the genres for reporting that research. Most McGill students were in the management faculty and were comfortable with analytical practices, like SWOT analysis and such genres as business plans. Four of the Delaware students were in the College of Business and Economics; many were nursing students, and others came from disparate disciplines. Aside from their experience in writing term papers for the freshman English course, they had little in common concerning research practices.

At the beginning of the semester, the Delaware students in particular found it difficult to conceive of a topic for analysis, lacking experience in reporting that would help shape their concept of how to do research toward a report. We thought our rather elaborate and detailed project Web site made the process and product abundantly clear; but without some intuitive or learned understanding of organizational research as it resulted in a document, the student teams had a difficult time negotiating a topic. The most successful teams were those with members who had had experience in a business discipline. That experience, however, had to translate into an openness to a new form of presentation—a comparative report—rather than stern adherence to another genre, for example, the business plan, the financial analysis, or the SWOT analysis.

The two most successful teams adhered closely to the genre specifications outlined in the assignment while also adapting them appropriately to reflect their increasing knowledge of their subject. With the other teams, we became aware of the downside of a highly explicit genre template intended to help students bridge different interpretations of the assignment in the two locations. To different degrees, the teams took a literalist view, which led them to create repetitive, excessively divided reports. In part, that may reflect our own well-intentioned if misguided effort to have one side set up the styles and headings in Word so that the teams had a consistent outline to fill in. In doing so, we may have imposed on the teams more than we aided them, and we tied our hands to some degree as we assessed their drafts.

But the teams did come to understand well the conventions of the milestone assignment genres. In addition, they easily adapted to the genre of the blog entry, with its more open and interactive aspects. The students' first project assignment was creating a team blog and contributing entries to introduce themselves to the rest of the team. As the semester progressed, messages became less oriented to personal narratives and included links to their universities, their towns, the companies they were studying, important documents they had found, and other relevant research and contextual information that kept them aware of what was happening at each site.

The blog's design immediately reinforced the sense of team: the opening page displayed a jointly written purpose statement for the blog, a logo and team name (for those teams that developed one), the names of the teammates, and an index to the messages archived on the blog, an automatically created team memory. As Röll (2004) notes, blogs make thought processes visible. The blog structure showed the interaction and connections between messages better than a mere series of e-mail postings. Students did complain, however, that blogs could be organized only chronologically, not in other ways that might have been helpful to them.

In class, we discussed the evolving norms of the blog, discussions that allowed us to show how genres change and become what they are in different contexts. In particular, we examined the "peculiar intersection of the public and private" (Miller & Shepherd, 2004) that blogs represent. They are part diary, part essay, part opinion piece. In their early blog postings, students were, by design, highly personal. Later, as they read corporate and other blogs, they became aware of how public blogs differ and began to question whether they should put their pictures on the blog, whether their research could be picked up by someone else, whether they'd run into trouble talking about Ben & Jerry's or Wal-Mart.

The first entry by an outsider on one of the team's blogs shattered a sense that obscurity and the sheer number of blogs protected their privacy. By the end of the semester, the blogs may have served their purpose of introducing team members and building a shared team space and archive, so that some teams and team members used their blogs only for casual postings. They found it easier to manipulate major assignments on ACollab or e-mail and shifted their attention from blogging to those assignments. For our part, we were not usually comfortable commenting *in* their blogs (although we had access to them), so our lack of comments there, too, may have sent a signal to students that they weren't important.

DEVELOPMENT OF COMMUNICATION
PROCESSES AND STRATEGIES

By the end of the class, most if not all students were proficient in navigating the genre conventions of business practices, such as structuring e-mail messages, blog entries, memos, and the shell of a professional report. Assessing students' abilities to edit for style, grammar, and mechanics turned out to be more challenging, although the students felt they had increased their proficiency here as well (see Figure 2).

It was hard for us as instructors to know where to set the bar on matters of style, grammar, and mechanics. Although we noted obvious problems in mechanics in our comments, our interest in seeing the project and the teams succeed in a very complex endeavor probably meant that we didn't pay as much attention as we might otherwise have to sentences and paragraphs in themselves. The need to motivate with praise sometimes trumped our need to point out areas for improvement. Those conflicting needs are obvious in any classroom, but this intercultural and virtual partnership presented some specific challenges.

One challenge was the sheer volume of writing the teams produced. We simply let some go without comment. Instead, we provided detailed written commentary on milestone assignments and talked with team members in general about what we found on their blogs. Another challenge was helping students develop an effective strategy and language for collaborative and virtual revising. In general, the teams settled into a pattern that resembled the model we used for our own commenting. Members in one location took the lead on a draft and posted the draft on ACollab or a blog. Those in the other location, sometimes mainly the team leader on the other side, then added comments. On the least effective teams, members at one site waited impatiently, assuming the other members were failing to do their part. Because the classes didn't meet at the same time (even though the two universities are in the same time zone and had roughly the same semester schedule), that asymmetry provided cause for ill feeling. The Delaware class met before the McGill one, on the same days (Tuesday and Thursday). On the early assignments, Delaware students wrote first and waited for McGill to catch

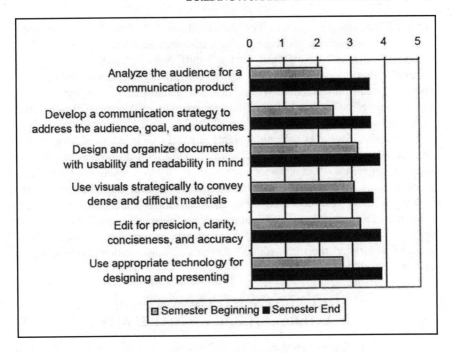

Figure 2. Students' self-rated proficiency in communication processes and strategies.

up. In time, the teams negotiated strategies to divide the work more evenly. We instructed students in what to look for in each others' writing, and we had backup material for them, including a textbook and pointers to online sources for help. They did not seem to incorporate much of the language for such review into their entries, however. Most comments focused on content (especially missing content) and on compliance with the structure of the assignment rather than on matters of expression. Sometimes, team members in one location even seemed to work against those in the other location, each creating their own text and requiring negotiation among competing segments and versions rather than working on edits within one version.

A third challenge was the varieties of English used by students, a result of the intercultural nature of the project. For many McGill students, English was a second (or third or fourth) language. The project also demonstrated the multiple conventions within English in a global context. One student wrote, for example, in the policies and procedures memo, that the final report would use "standard Canadian English." In the end, in part because of an accident of logistics (the Delaware semester lasted longer than the McGill one, and thus the Delaware side had the last review before submitting the final report), U.S. English as enforced by Word's spellchecker prevailed. But within the reports there were some reminders of differences: in measurements (metric vs. English), in diction ("surnames" versus "last names," "bill" versus "check," "gas" versus "petrol," "metro" versus "subway," and so on). English speakers in Québec have also included more French words in their language, for example taking the "autoroute" rather than the "highway" or going to the "depanneur" rather than the corner store.

A final challenge was making students aware of their authorial identity and situatedness as they wrote. As they were working with teammates in different locations to create documents for audiences in different locations (sometimes us as teachers, sometimes their teammates, and sometimes the managers of the businesses they were analyzing), the students realized that phrases referring to their location or national affiliation such as "here in Canada" did not work well in an intercultural context. Instead, they had to learn how to expand their perspective beyond their immediate location and think of their situatedness in an intercultural context.

THE ABILITY TO COMMUNICATE
IN INTERCULTURAL VIRTUAL TEAMS

To the extent that all teams completed all assignments, by definition, they had learned to communicate in an intercultural and virtual context, one they had never worked in before. The end-of-semester survey also reflected their increased proficiency in such work (Figure 3).

The project was an immersion experience for students, similar to learning a language through living in a culture rather than just studying rules and vocabulary. They learned the technology for collaboration through using the technology, and they learned intercultural team dynamics through participating on an intercultural team. Such learning was hardly risk free, and our highly detailed assignments notwithstanding, it had its moments of muddle and uncertainty. One early moment was signaled by an ACollab message headed, "DELAWARE: WE HAVE A PROBLEM." Although the Delaware members of the team at first thought this meant they had caused the problem, we pointed out the use of "we," which happily signaled a collaborative approach; it wasn't "you" have a problem.

In their research on the business communication practices of multinational companies operating in two different local contexts, students learned how context—local, national, industrial—influences practices and communication choices. Students learned that despite the need to protect and represent a unified brand, successful multinationals adjust to the pulse of their local communities and the practices of their local competition. The Starbucks team, for example, discovered different customer preferences and methods of communicating with customers, different peak seasons and peak business times, different organizational structures and levels of formality in communicating within the company,

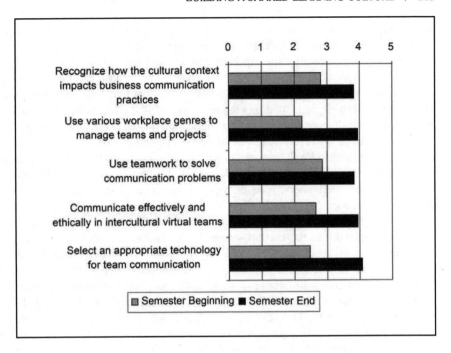

Figure 3. Students' self-rated proficiency in intercultural
virtual team communication.

and different forms of employee recognition and thus attitudes toward motivating employees. Similarly, the IKEA team uncovered the impact of
local holidays on peak business seasons. For example, in Québec, the twin
holidays of the province on June 24 and of Canada itself on July 1 has caused
the first of July to be the customary day for anyone who is thinking of moving
to move, and the days following July 1 have become the furniture store's peak
business period.

They also learned how genre conventions can help span their own differences
in context to create, in effect, a unified brand in their work, both in terms of their
working processes and of the final communication product. Their policy memo,
for example, helped them foresee potential difficulties in team dynamics. The
project-planning proposal forced them to define a shared goal, contextualize their
research together, clarify their methods, and coordinate their research and writing.
The most successful teams went beyond the mere specifications of the assignment
genres to bond as a team, something shown neatly in their oral presentations,
which students gave in their respective locations. Culturally integrated teams
presented the names and photos of all teammates on their first slides, while
others did not mention their teammates in the other location at all or discussed

their work with their teammates only toward the end of their presentation. Even in their last blog postings, the McDels, for example, used the red and blue team name to thank each other for the experience of working together.

Around the milestone genre assignments, students built those bonds, or failed to do so, through rich and extensive electronic conversations—on ACollab, on their blogs, in e-mail, in instant messaging. Some of that conversation centered on how they would converse—metacommunication that did not come naturally. They recognized, however, that virtual work demanded that other layer of message. They had to build mutual knowledge, of course, but they also had to build rapport, and these conversational gambits did that. Team leaders in particular became cheerleaders, recognizing and appreciating the work of teammates and of encouraging each other to greater efforts. At times, 75% of a blog posting would consist of such metacommunication. The tone was often relentlessly upbeat and the diction informal. The entries for all-female teams, for example, consistently used the expression "girls," one that we found a bit off-putting but they saw as appropriate. The blogs resounded with comments like these: "it has been a pleasure working with you girls. I think our team did a great job!" or "Thanks for all your hard work thus far!" or "Keep up the good work girls—GO TEAM GO!" or "good job on finding links about Marriott Newark." On the blogs, students also exchanged wishes for national holidays (e.g. Thanksgiving), discussed their holiday experiences, and even included the drawing of a turkey ("drawn" with various symbols such as slashes and parentheses).

On the two least successful teams, early conflicts over topics and genres hindered performance throughout the term. It is interesting to speculate about the role cultural differences may have played on these two teams. On one team, strong personalities, perhaps derived in part from ethnic origins and gender expectations between the team leader and vice team leader, contributed to the often contentious behavior or faultlines between the two locations. Even at the beginning, students encountered difficulties negotiating the communication practices through which they would agree on such things as a topic and the use of generic conventions for their situation. While the team leader (a woman) at Delaware argued for inclusive communication practices, the vice team leader (a man) at McGill advocated a more hierarchical approach for the sake of efficiency, in which he and the team leader would discuss issues and communicate their decisions to teammates in each location. Indeed, throughout the semester, the team experienced a number of the vulnerabilities described by Cramton (2002) and Cramton and Hinds (2004). For example, when scheduling difficulties arose in arranging a synchronous meeting (a chat), the team leader attempted to facilitate decision making by suggesting that decisions made during the chat be final and that those who did not attend the chat simply accept the decisions without any further input. This suggestion was rejected and criticized as "harsh" by a McGill teammate. Apparently, communication of contextual information had been insufficient to fully appreciate the work practices and schedules of teammates in both locations.

The second-least successful team was the only one with an international student in the Delaware location. Relations between team members and that student, as reported in the blogs and in a face-to-face conversation with the student, were often testy. She accused the team of ignoring her contribution; the team accused her of misunderstanding what she was to contribute. At every class, she spoke directly with the instructor rather than the team leader in asking questions and in other ways seemed to look to the instructor as authority figure in the classroom, rather than her team leader, perhaps reflecting her expectations of learning as shaped in her home institution.

Outcomes for Faculty

One of the main benefits of such international partnerships is the opportunity it affords faculty to converse about both course content and pedagogy in a global rather than local institutional context. The partnership thus presented a unique opportunity for us to talk through our pedagogical decisions, our use of technologies in the classroom, our teaching philosophies, and more. In addition, the project prompted us to discuss our institutional cultures and how they impact our teaching and assessment practices. The project encouraged us to create a collaborative instructional approach that neither of us could have created alone in her classroom.

In developing such a partnership, both instructors need to work with similar courses, ones that place similar emphasis, for example, on teamwork, on the technology for collaboration, on project-based learning, and on cross-cultural understanding. A real commitment to working internationally is also important to sustain interest across the challenges such work presents. In addition, it helps if the calendars of the partner institutions are similar. To the degree that the calendars differ, the courses occur in different time zones and seasons, and the language proficiency of students at the partner institutions differ, negotiations about logistics and approaches becomes more complex.

Conducting such negotiations virtually immerses the partnering faculty in the same issues their students face, which is not a bad condition in which to assess those students' work. We were fortunate in being equally attentive to e-mail and focused on the project. At the beginning of the semester, our messages focused largely on glitches and opportunities in the technologies, which we, too, were learning with our students. Later, we spent more time discussing adjustments in the assignments and deadlines, comments we should be making on the drafts and final reports, grading, and strategies for aiding underperforming teams. From time to time, one of us served as an advocate for teams that were having trouble meeting the deadlines as established.

The project also allowed us to learn a lot from our students as they collaborated and interpreted our instructions or classroom comments with each other in order to coordinate their work. Hearing and seeing how they had interpreted our comments in each location helped us clarify and restate them to better support the students in their team-building and writing efforts. Most important, we recognized the need to continually emphasize a shared team culture in our teaching and in our assessment so that students could move beyond the perhaps natural faultlines of their physical locations.

Finally, the project showed the importance of a solid partnership behind a project that spans institutional, cultural, and national boundaries and thus requires a substantial amount of sharing, negotiating, and adjusting of practices. As Duin and Starke-Meyerring (2003) point out, the process of building such a learning partnership is complex. All we had at the beginning was a shared goal and a strong commitment to providing our students with an immersion experience that would facilitate their ability to learn how to communicate effectively in an intercultural and digitally mediated context. The blueprint Duin and Starke-Meyerring provide for partnering then served as a heuristic for the questions we would need to work through, issues we would need to articulate, and the things we would need to put in place.

CONCLUSIONS AND RECOMMENDATIONS FOR ASSESSMENT OF INTERCULTURAL VIRTUAL TEAMS

The virtual team project between our two management communication classes provided, for us as well as our students, a rich intercultural learning environment and immersion experience. But such a culture of learning does not develop automatically or without problems. Simply bringing two or more classes together via the Internet does not suffice. We had to build a culture and maintain it by using consistent and meaningful assessment. In doing so, we have extracted these preliminary lessons:

- Throughout the project, we found that our focus on the main goal of the project—helping students learn to build a shared virtual learning culture across locations—was critical to the success of the project, as this shared learning culture both thrived on and enabled the negotiation of different practices and assumptions, and therefore ultimately made the sharing and making of knowledge across these differences possible, thus facilitating the best contributions from all team members.
- Given the objective of building a shared virtual-learning culture and its gradual process nature, a constructivist approach employing formative, dynamic, and interactive assessment strategies turned out to be important in making these outcomes possible. Interactive assessment helped students develop next steps, encouraged joint reflection during revisions, and

facilitated the development of team dynamics. In addition, it provided us as instructors with feedback for making decisions about teaching—about how to best support the work of the teams and the learning that occurred in them.

- As an immersion experience, intercultural virtual teamwork is particularly rich in teachable moments, as students struggle to motivate distant team-mates or rejoice in shared successes. Our assessment strategies became especially useful when taking advantage of this real-life intercultural engagement.

- Comments should reinforce the development of team dynamics. Often, the different learning outcomes intersect. For example, editing problems may surface as divisive language that does not reflect the situatedness of authors and audiences in an intercultural context (e.g. "here in Canada," or "Both the Delaware team and the McGill team investigating the . . .").

- Throughout a virtual team project, equal treatment is crucial to preventing faultlines (Cramton & Hinds, 2005) from erupting. We tried to foster such treatment through collaborative assessment and similar percentages for final grading. A problem, however, as in any team project, is addressing the reality of slacker team members. In some ways, our assessment of individual blog entries tried to account for this, but the approach probably was not enough, and some resentment remained. An individual learning-log assignment might help to overcome this problem.

- Finally, instructors should negotiate different assessment practices across institutions and develop a shared approach to assessment that models a shared learning culture for students.

Overall, the intercultural virtual team project provided students and us with an important, enjoyable, and rewarding immersion in intercultural learning and communication. Assessment approaches, choices, and practices played a key role in our work.

APPENDIX A
Point System for Grading

Criteria	Points
Context (Rhetorical Situation)	
Content	
Organization	
Language Use and Style	
Design and Visual Presentation	
Grade	

APPENDIX B
Pre- and Postproject Survey
(Where I stand as a professional communicator)

1. Please note your current proficiency in each of these genres on a scale of 1–5, where 1 = low proficiency, 3 = moderate proficiency and 5 = high proficiency:

Memo
Letter
E-mail
Formal report
Progress report
Proposal
Blog (Web log)
Team oral presentation
PPT presentation
Abstract or executive summary
Policies and guidelines
Interviewing and being interviewed

2. Please note your proficiency in each of these communication strategies in a management environment on a scale of 1–5, where 1 = low proficiency, 3 = moderate proficiency and 5 = high proficiency:

Analyze the audience for a communication product.

Develop a communication strategy to address the audience, goal, and outcomes of a communication situation.

Design and organize your documents with usability and readability in mind.

Edit for precision, clarity, conciseness, and accuracy.

Use visuals strategically to convey dense and difficult material.

Use appropriate technology for designing and presenting information and choosing the right media for the situation.

3. Please note your proficiency in each of these elements of teamwork on a scale of 1–5, where 1 = low proficiency, 3 = moderate proficiency and 5 = high proficiency:

Use various workplace genres, including proposals, memos, and reports, to manage teams and projects.

Recognize how the cultural context impacts business communication practices.

Manage a fairly complex professional research and reporting project.

Use teamwork to solve communication problems.

Communicate effectively and ethically in intercultural virtual teams.

Select an appropriate technology for team communication.

REFERENCES

Andrews, D. C., & Andrews, W. D. (2004). *Management communication: A guide.* Boston, MA: Houghton Mifflin.

Cohen, S. G., & Bailey, D. E. (1997). What makes teams work: Group effectiveness research from the shop floor to the executive suite. *Journal of Management, 23,* 239–290.

Cramton, C. D. (2002). Finding common ground in dispersed collaboration. *Organizational Dynamics, 30*(4), 356–367.

Cramton, C. D., & Hinds, P. J. (2005). Sub group dynamics in internationally distributed teams: Ethnocentrism or cross-national learning? In B. M. Staw & R. M. Kramer (Eds.), *Research in organizational behavior, 26,* (pp. 231–263). Greenwich, CT: JAI Press.

Duin, A. H., & Starke-Meyerring, D. (2003). Professional communication in the learning marketspace: A call for partnering. *Journal of Business and Technical Communication, 17*(3), 346–361.

Earley, P. C., & Mosakowski, E. (2000). Creating hybrid team cultures: An empirical test of transnational team functioning. *Academy of Management Journal, 43*(1), 26–49.

Herrington, T. (2004). Where in the world is the global classroom project? In J. Di Leo & W. Jacobs (Eds.), *If classrooms matter: Progressive visions of educational environments* (pp. 197–210). New York: Routledge.

Herrington, T., & Tretyakov, Y. (2005). The global classroom project: Troublemaking and troubleshooting. In K. Cargile Cook & K. Grant-Davie (Eds.), *Online education: Global questions, local answers* (pp. 267–283). Amityville, NY: Baywood.

Keller, J. (2007, September 19). Canadian universities find ways to keep private info from U.S. Patriot Act. CBC News. Retrieved September 19, 2007, from http://www.cbc.ca/cp/technology/070919/z091921A.html

Miller, C. R., & Shepherd, D. (2004). Blogging as social action: A genre analysis of the Weblog. In L. J. Gurak, S. Antonijevic, L. Johnson, C. Ratliff, & J. Reyman (Eds.), *Into the blogosphere: Rhetoric, community, and culture of Weblogs.* Retrieved December 28, 2004, from http://blog.lib.umn.edu/blogosphere

Ramsden, P. (1992). *Learning to teach in higher education.* New York: Routledge.

Röll, M. (2004, July). *Distributed KM: Improving knowledge workers' productivity and organisational knowledge sharing with Weblog-based personal publishing.* Paper presented at BlogTalk 2.0, Vienna, Austria. Retrieved December 28, 2004, from http://www.roell.net/publikationen/distributedkm.shtml

Sapp, D. (2004). Global partnerships in business communication. *Business Communication Quarterly, 67,* 267–280.

Scott, B., Longo, B., & Wills, K. (2006). Introduction: Why cultural studies? Expanding technical communication's critical toolbox. In B. Scott, B. Longo, & K. Wills (Eds.), *Critical power tools: Technical communication and cultural studies* (pp. 1–19). Albany, NY: SUNY Press.

Selber, S. (2004). *Multiliteracies for a digital age.* Carbondale, IL: Southern Illinois University Press

Shepard, L. (2000). The role of assessment in a learning culture. *Educational Researcher, 29*(7), 4–14.

220 / ASSESSMENT IN TECHNICAL AND PROFESSIONAL COMMUNICATION

Starke-Meyerring, D., & Andrews, D. (2006). Developing a shared virtual learning culture: An international classroom partnership. *Business Communication Quarterly, 69*, 25–49.

Starke-Meyerring, D., Duin, A. H., & Palvetzian, T. (2007). Global partnerships: Positioning technical communication programs in the context of globalization. *Technical Communication Quarterly, 16*(2), 139–174.

Walvoord, B. E., & Johnson Anderson, V. (1998). *Effective grading: A tool for learning and assessment.* San Francisco, CA: Jossey-Bass.

Zhu, Y., Gareis, E., O'Keefe Bazzani, J., & Rolland, D. (2005). A collaborative online project between New Zealand and New York. *Business Communication Quarterly, 68*(1), 81-96.

CHAPTER 14

Do Fish Know They Are
Swimming in Water?

Deborah S. Bosley
University of North Carolina, Charlotte

In "Assessment in an Intercultural Virtual Team Project: Building a Culture of Intercultural Learning" (see Chapter 13, this volume), Starke-Meyerring and Andrews have presented us with a much needed comprehensive description of the processes of designing virtual teams and the methods for assessing changes in student behavior and knowledge. Quantifying student experiences and learning give us one view of how they are affected by virtual team assignments. However, the voices of the students themselves are another important means of assessing the successes of such course linkages. This response to their article is derived from assessment comments from my own students who were involved in a similar assignment, though less fully formed and prepared than the Starke-Meyerring and Andrews's work. First, I will describe, in general terms, the conditions of a similar collaboration with my colleague Lucy Veisblat of the Université Paris 7—Denis Diderot. Then, I will let the students "speak" for themselves.

DESCRIPTION OF THE COURSES AND
THE ASSIGNMENT

Our Paris partners were enrolled in *Introduction to Technical Writing: Writing International Documentation*. The Paris 7 group consisted of 20 graduate students, in a technical writing cooperative education experience, alternating three weeks full-time at the university taking courses with three weeks full-time in their company. All were native French speakers, except for one native English speaker and one native German speaker, and all were fluent in English. Most had

221

a degree in foreign languages and translation. My course, *Introduction to Technical Communication*, enrolled 14 English graduate students, almost all of whom were working either part-time or full-time while taking one to three graduate courses. Despite the differences in numbers of students, my Paris colleague and I decided to have the same number of teams, with the U.S. students working in pairs and the Paris students in teams of 3 to 4. We linked our student teams and assigned them to write a set of instructions for cell phones for an audience of new cell phone users. Their task was to edit the set of instructions created by their virtual team members. My counterpart in Paris and I began our assignment by having our students exchange brief biographies. The student biographies generated questions, comments, surprises, and curiosity about their virtual teammates. As one student commented, "Reading their self descriptions was in some ways motivational for me; it personalized the experience and helped lower a 'personal resistance barrier' that I would have otherwise sustained throughout the project."

Each team then wrote rough drafts of their cell phone instructions, e-mailed them between the two peer-review teams, revised according to suggestions by each team, conducted user testing at their home sites, and turned in their final copy. At the conclusion of the assignment, the U.S. students wrote individual memos reflecting on their collaborative experience. Many of these comments indicate a level of curiosity that I rarely see; they were genuinely intrigued by the idea of communicating with students outside of the United States. Their comments tended to group around the three outcomes that Starke-Meyerring and Andrews describe: (1) developing proficiency in written and oral professional communication genres; (2) mastering communication processes and strategies; and (3) communicating successfully in intercultural virtual teams.

COMMENTS ABOUT THE PROCESS AND THE COLLABORATION

As was true with Starke-Meyerring and Andrews's experience, none of my students had worked on an intercultural, virtual team. Again, as with the Starke-Meyerring and Andrews project, working across the globe presented a number of collaborative challenges for our students:

> Often in a collaborative environment there is an uneasy harmony of ideas, which may indicate that all members of the group are not applying themselves to the task at hand as they should. Online collaboration is inherently impersonal, allowing members to be more open to giving and receiving criticism. Constructive criticism should be encouraged because the impersonal nature of the medium is well-suited for it. On the other hand, care should be taken to avoid conflict online as conflict resolution [online] becomes difficult.

Despite the general success of this international collaboration, students articulated a concern for the lack of personal interaction with our Paris partners and the "bonding," or lack thereof, that can occur with virtual teamwork:

> Because of this collaboration experience with students in Paris, I learned that collaboration through email might be efficient, but lacks the personal, one-on-one interaction that adds so much to my own writing process. I found my experience with my in-class partner to be more exciting and more enlightening that my experience with the Paris partners. Since the audience is such an important consideration for a technical writer, I think that personal interaction with a collaboration partner provides not only another technical writer's perspective, but also the perspective of a one-person audience. That is what I missed with the Paris partners and what I learned is essential in my own process.

COMMENTS ABOUT LANGUAGE AND WRITING

As Starke-Meyerring and Andrews note, students "had to negotiate a number of different preferences, practices, and interpretations of assignments." My students were quite articulate about the language and writing requirements of this assignment:

- The comments that I received back from the French students were very helpful. We noticed that they made their editing remarks in question form. For instance, they wrote, "This sentence is not related to the fact of getting the new phone so wouldn't you prefer to put it in another section as a comment?" An American student would have simply written "bullet point," or crossed out the sentence and drawn an arrow to indicate moving the sentence to a more appropriate section.
- We found that even though the Paris partners communicated in English, some of their English words and phrases were confusing to us. For instance, in one section of their instructions, the command was to "load the battery" and a bulleted item discussed "loading for the first time." My partner and I thought this might mean "charge the battery." In our comments to the Paris partners, we mentioned that we thought "load" meant "charge," which they agreed to.
- I realized that I wanted to use the same tone and level of familiarity that [one of my Paris partners] used when she wrote to us, so that I did not offend . . . in some way. I learned that I wanted to be careful to match the email style of our international partners.

COMMENTS ABOUT DIFFERENCES/SIMILARITIES

The U.S. students also were quite perceptive about what they assumed to be differences and similarities in cultural expectations:

- I saw several cultural differences that pointed out how isolated most Americans are compared to Europeans. Our French partners were all bilingual or multilingual, whereas most Americans speak only English. Our partners were also very well traveled and had lived in other countries. I have always lived in North Carolina.
- Another difference was in the tone of their communication. Although they took this project seriously, their communications with us were very light-hearted. Their emails were very casual, much more so than I would normally write to strangers and to work partners. It was charming, but surprising to me.

FINAL COMMENTS

Like fish who do not know they swim in water, our American students often do not realize that their culture is not the culture of the world. That they need help to recognize "the water" is best illustrated in this comment one American student made near the beginning of the project: "It did not occur to me that our Paris partners would not have a holiday on November 25, Thanksgiving . . . It took me a while to realize that not everyone celebrates Thanksgiving. Of course this makes sense, but it did take me a second to shift my thinking."

Students' understanding of all the complexities of stretching beyond their own culture benefits immeasurably from virtual team projects, like the one so fully described in the Starke-Meyerring and Andrews's article. Globalization is occurring at an astounding pace, and we want our students to participate fully as world citizens. Accomplishing this outcome means having the tools for assessment—assessing how well our students understand themselves and others as the critical component to enable us to help them succeed in our new world.

The Ethical Role of the Technical Communicator in Assessment, Dialogue, and the Centrality of Humanity

Sam Dragga, Texas Tech University

The Foreword to this volume makes a vigorous case for "meaningful actions" in assessment, but without a detailed discussion of the influence of ethics in this process. It is thus my objective in this Afterword to focus on ethics.

If we conceptualize ethical judgments (including judgments regarding both verbal and visual communication) along two intersecting lines—from absolute to relative and from universal to particular—we would perceive four possible judgments: all people should always do X (absolute-universal), some people should always do X (absolute-particular), all people should sometimes do X (relative-universal), and some people should sometimes do X (relative-particular). Figure 1 illustrates the four judgments.

The information age changes radically the practice of ethics in the creation of both verbal and visual information, making it critical to deliver the right information at the right time. In a world of always-growing and changing information, technical communicators must be filters—ethical filters—instead of containers, picking their way through the available information to find genuine knowledge. Effective communication—ethical communication—allows us to make knowledge from information and (sooner or later) to derive wisdom from this knowledge. Prevailing theories of ethics, however, operate on container metaphors and encourage thinking about the stability of data, the shaping of information, and the durability of practice instead of the fluidity of data, the sifting of information, and the dynamics of practice.

In the classical ethics of Plato and Aristotle, for example, the emphasis is on the character of the individual and his or her virtues and vices: that is, virtuous

225

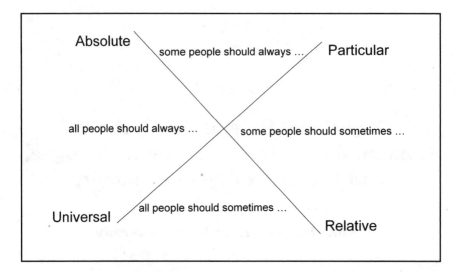

Figure 1. Four kinds of ethical judgments.

people know the right thing to do. And rationalist ethics, typically focusing either on obligations that influence behavior or on the consequences of behavior, emphasizes the making of logical decisions; that is, rational people know the right thing to do. Both classical ethics and rationalist ethics operate on a "container" metaphor: the soul must be filled with virtues and emptied of vices, or the mind must be filled with logic and emptied of emotion.

Obviously, different professions have different perspectives on ethics, influenced by the specific issues and practices of each profession. Technical communicators, for example, are likely to espouse a particular-relativist position on ethics because we have developed the habit of adapting information (and its organization, style, and design) to the different people using that information. While the engineer designing a machine typically perceives all users of that machine as essentially identical, technical communicators try to be sensitive to the individual variations that might influence a user's interpretation of the operating instructions.

Each of us, however, has multiple perspectives (personal, religious, political, economic, cultural, etc.) and multiple obligations (to family, organization, profession, society, etc.) and slip readily from universal and absolutist principles (considering people identically) to particular and relativist principles (considering people individually).

While it is tempting, for example, to consider all the people within a company, profession, or society identically (using universal and absolutist principles), to do so is to ignore their individuality and to ignore the relationship we have with each human being as a human being. In doing so, we diminish the individual's humanity, perceiving him or her with rigorous indifference. We essentially declare this unique individual unexceptional: we thus seem uncaring, insensitive, inflexible. Nevertheless, if people in a company, profession, or society are to function effectively, a degree of subordination of individuality (and humanity) is necessary: all of us must yield to cooperative principles of operation. Otherwise, we seem inconsistent, erratic, unfair. We are thus continually shifting from universal-absolutist to particular-relativist perspectives, considering people as identical across time and as individuals in time.

Dialogue is a way to negotiate this daily collision of absolutist versus relativist and universal versus particular perspectives (as well as private/personal versus public/professional perspectives). The trick is negotiating ethically if people possess unequal power, money, education, experience, sensitivity, flexibility, and the like.

If technical communicators are going to be effective and ethical filters of information, we might look especially to ethical cases that immerse us in the messiness of volatile, amorphous, and transitory life, and make us recognize that the principles we consider absolute and universal always occur within a situated narrative and might thus be only relative and particular. In the application of a theory of ethics (a container), we discover the limits of the theory and the necessity of adapting/interpreting the theory to the variables of life (making of it a filter). The danger in using cases, however, is that we come to be addicted to the logical and deliberative process of adaptation and interpretation. We come to imagine that we'll always have sufficient information, time, and insight to decide every ethical dilemma. Ordinarily, we won't. Nevertheless, such stories illuminate the possibilities for virtue and vice, for rational and irrational thinking, for right and wrong behavior in a given situation, from which we might derive principles that are either absolute-particular or relative-universal.

And as we discuss the social responsibility of technical communicators and the limits of that responsibility, consider this possibility: in the development and documentation of products and services, managers privilege profits and productivity, and engineers privilege effectiveness and efficiency. It is thus the job of technical communicators to privilege humanity, to serve as the voice of the real human beings who will use the product or service or otherwise be influenced, directly or indirectly, by the organization's operations (e.g., through its environmental or labor practices).

I don't wish to suggest that the considerations are always of equal importance or always opposed to each other. The impact of a decision on people might be minor, but yield sizable profits. Or, what's good for efficiency might also be good for people.

While managers and engineers might from time to time weigh the humanity of their decisions (as technical communicators must weigh the impact of their decisions on profits, productivity, effectiveness, and efficiency), humanity is often their lower priority, their lesser consideration. It is the technical communicator's job, using both words and illustrations, to keep the human impact of products and services and of policies and practices a prominent factor in the organization's dialogue. I don't think that ethical and professional technical communicators always have to dominate that dialogue or necessarily prevail in their opinions: I do think we must always raise the issues and ask the questions so that genuine dialogue occurs. We all have to avoid excesses of objectivity and subjectivity (and both passivity and self-righteousness), and a continuing dialogue is a good vehicle for keeping us all—managers, engineers, and technical communicators— receptive and responsive to each other as well as to the societies in which we live.

As we proceed with assessment initiatives, let us keep in mind that in doing so, we are guardians of the ethics of the discipline. In addition to all that we might assess, it is thus imperative to include the ability of programs to teach technical communicators to (1) serve as filters of information instead of containers, (2) to establish dialogic relationships with all professions, and (3) to give voice to humanity in both verbal and visual communication.

And as we proceed with assessment initiatives, let us lead by example— filtering information to generate knowledge and wisdom about successes and failures, enjoying dialogue with colleagues across the disciplines, and amplifying the voice of humanity in both teaching and learning.

Contributors

JO ALLEN is Senior Vice President and Provost at Widener University in Chester, Pennsylvania, where she oversees eight schools and colleges, student affairs, the libraries, academic support services, and special programming. Before her work at Widener, she was a faculty member and administrator at North Carolina State University, where she helped promote student learning outcomes assessment as part of her role as Associate Vice Provost for Undergraduate Academic Affairs. Earlier in her career, she served as a faculty member and administrator at East Carolina University, where she co-directed programs in technical and professional communication and supported the assessment work from the Office of the Vice Chancellor for Academic Affairs.

PAUL ANDERSON is Director of the Roger and Joyce Howe Center for Writing Excellence and Professor of Technical and Scientific Communication at Miami University, Ohio. He served on Miami's Assessment Task Force, chaired its Accreditation Steering Committee, and served on its strategic planning council. He has been Special Assistant to three provosts and chaired the University Senate's Executive Committee. In his current research, he employs quantitative methods to study writing and learning in higher education. Dr. Anderson believes that examination of student writing is the best way for universities to assess their effectiveness at teaching critical thinking and other higher-order mental activities.

DEBORAH C. ANDREWS, Professor of English at the University of Delaware, coordinates the department's concentration in professional writing and directs the university's Center for Material Culture Studies. She teaches courses in technical editing, in researching and interpreting objects and sites for public understanding, and in American literature from 1865 to 1945. She has published several texts on professional communication, including *Technical Communication in the Global Community* and, most recently, *Management Communication: A Guide*. A researcher, consultant, and speaker on many aspects of professional communication, especially in an international context, Dr. Andrews is the former editor of *Business Communication Quarterly*.

CHRIS ANSON is University Distinguished Professor of English and Director of the Campus Writing and Speaking Program at North Carolina State University, where he helps faculty in nine colleges to use writing and speaking in the service of students' learning and improved communication. Before joining NC State in 1999, he spent 15 years at the University of Minnesota, where he directed the Program in Composition (1988-1996) and was Morse-Alumni Distinguished Teaching Professor. His research interests include communication across the curriculum, writing to learn, response to writing, and assessment, topics on which he has published and spoken widely.

DEBORAH S. BOSLEY, Director of the Center for Humanities, Technology, and Science at the University of North Carolina at Charlotte, is Associate Professor of English and teaches technical communication to undergraduates and graduate students. Her research focuses on technical communication to improve comprehension of written information in the workplace. She has conducted numerous usability tests for nonprofits and corporate entities, as well as teaching usability and assessment courses at the graduate and undergraduate levels. Dr. Bosley has published more than two dozen articles and three books, including *Global Contexts: Case Studies in International Technical Communication*. She has spoken at international conferences in Mexico, France, Ireland, Germany, Spain, and England on usability testing and other means of assessing the efficacy of written information.

MICHAEL CARTER is Professor of English and Associate Dean of the Graduate School at North Carolina State University, where he has taught a wide variety of courses in writing and rhetoric, including undergraduate and graduate courses in writing in the empirical sciences. He is the author of *Where Writing Begins: A Postmodern Reconstruction* and articles in such journals as *College English*, *College Composition and Communication*, and *Research in the Teaching of English*. He received the 2008 Richard Braddock Award. Dr. Carter has taken a major leadership role on his campus in the assessment of general education and undergraduate and graduate academic programs.

KELLI CARGILE COOK is Associate Professor of Professional and Technical Writing at Utah State University, where she teaches both graduate and undergraduate courses. She chaired the undergraduate program at Utah State for 3 years and now leads the doctoral program in Theory and Practice of Professional Communication. She has served on the executive board of both the Council for Programs in Technical and Scientific Communication and the Association for Teachers of Technical Writing. Dr. Cook's research interests include online education, technical communication program development, onsite and online course assessment, web-based training, and cybercultures.

NANCY W. COPPOLA is Professor of English at New Jersey Institute of Technology, where she directs the graduate program in professional and technical communication. She teaches graduate courses in technical communication, technology transfer, and usability issues. Her research on program assessment, technology transfer, research methods, and environmental rhetoric has been widely published in journals, books, and edited collections. Dr. Coppola serves on the task force to create the STC Body of Knowledge and is co-founder of the CPTSC Research Assessment portal, both significant efforts to provide valid structure and methodology for technical communication assessment.

SAM DRAGGA is Professor of Technical Communication and Chair of the Department of English at Texas Tech University. He is a Fellow and former President of the Association of Teachers of Technical Writing and a recipient of the Society for Technical Communication's Jay R. Gould Award for Excellence in Teaching Technical Communication. He is co-author of *Reporting Technical Information* (11th ed., Oxford University Press, 2006) and is series editor of Allyn & Bacon's *Series in Technical Communication* (19 titles). He has also published a variety of journal articles on ethics in technical communication.

JAMES M. DUBINSKY is Associate Professor of English at Virginia Tech; for the past ten years, he has directed the Professional Writing Program, a program he was hired to build. He is a recent winner of a college award for outreach and the university's teacher scholar award. His research focuses on community-university partnerships, assessment, and pedagogy. Dr. Dubinksy is the author/editor of *Teaching Technical Communication*; he has contributed to journals such as the *Michigan Journal of Community Service Learning*, and edited an issue of *TCQ* on civic engagement. He is also vice-chair of the board for the YMCA at Virginia Tech.

NORBERT ELLIOT is Professor of English at New Jersey Institute of Technology. Holding the designation of Master Teacher, he offers courses at both the undergraduate and graduate levels in the areas of research methods, communication theory, corporate communication, and editing. A researcher in writing assessment, he is author of *On a Scale: A Social History of Writing Assessment*, a volume that was awarded the 2007 Outstanding Book Award from the Conference on College Composition and Communication. Recent work has appeared in *Technical Communication, Journal of Writing Assessment*, and *Rhetoric Review*. The centrality of Dr. Elliot's place in the field was marked by an invitation to write *The Enduring Vision of Henry Chauncey*, a monograph celebrating the sixtieth anniversary of the founding of the Educational Testing Service.

BILL HART-DAVIDSON is Co-Director of the Writing in Digital Environments Research Center at Michigan State University. He is Associate Professor of Rhetoric and Writing, teaching graduate and undergraduate courses in technical and professional writing. His research is focused on creating information systems and technologies that represent theoretically sound interventions in knowledge work and organizational writing. Assessment of writing becomes a critical lens for such work, as writing practices provide the most reliable index of organizational and personal goals in knowledge work settings.

232 / ASSESSMENT IN TECHNICAL AND PROFESSIONAL COMMUNICATION

MARGARET HUNDLEBY teaches in the Engineering Communication Program at the University of Toronto, after working in programs in the United States and elsewhere in Canada. She has also been an industry consultant in Canada, the United States, and the United Kingdom. She approaches technical communication work through rhetoric and the social organization of knowledge, centering her research on epistemic aspects of technical communication and the relations between written text and visuals. Dr. Hundleby has been part of several major assessment undertakings at the University of Guelph (Ontario), Michigan Tech, and Auburn University, and she served as founding liaison for the first CPTSC/ATTW assessment initiative. She was also a member of the Engineering Assessment Consortium, meeting at Rice University.

JEFFREY JABLONSKI is Associate Professor of English and Assistant Director of General Education at the University of Nevada, Las Vegas. He earned his Ph.D. from Purdue University in 2000. His research interests include professional/technical writing, writing across the curriculum (WAC), and writing program administration. His book *Academic Writing Consulting and WAC: Methods and Models for Guiding Cross-Curricular Literacy Work* (Hampton Press, 2006) examines how writing specialists collaborate with non-writing specialists to promote cross-curricular literacy in academic contexts. Dr. Jablonski teaches courses in composition theory, writing for the World Wide Web, and business and technical writing. He is also the book review editor for the *Journal of Business and Technical Communication*.

ED NAGELHOUT is Director of Business Writing at the University of Nevada, Las Vegas. He has edited two scholarly collections, published 15 articles or chapters, and presented more than 60 papers on professional writing, writing assessment, and writing program administration. He is the co-editor of the *ATTW Bulletin* and has served as a Stage I Reviewer for CCCC.

GERALD SAVAGE teaches technical communication and rhetoric in the English Department at Illinois State University. His research has focused on issues of professionalization in technical communication. Currently he is studying social justice issues for international technical communication. He has worked with accreditation self-study team for Illinois State University and has served as an external reviewer for program reviews at other universities. He co-edited *Writing a Professional Life: Stories of Technical Communicators On and Off the Job* with Dale Sullivan and the two-volume collection. *Power and Legitimacy in Technical Communication* with Teresa Kynell Hunt.

DOREEN STARKE-MEYERRING is Assistant Professor of Rhetoric and Writing Studies in the Department of Integrated Studies in Education at McGill University in Montreal, where she co-directs the Centre for the Study and Teaching of Writing. Her teaching and scholarship focus on writing and discourse in academic, workplace, and public contexts, especially as these contexts undergo change as a result of emerging digital technologies and globalization. She is the co-editor of *Designing Globally Networked Learning Environments* (Sense

Publishers, 2008), *Research Communication in the Social and Human Sciences* (Cambridge Scholars Publishing, 2008), and *Writing (in) the Knowledge Society* (Parlor Press/WAC Clearinghouse, forthcoming).

STEVEN YOURA directs the Hixon Writing Center at the California Institute of Technology, where he teaches courses in science communication and works with writing across the curriculum. Before coming to Caltech in 2001, he spent 14 years at Cornell University, where he was founding director of the Engineering Communications Program. He has published articles and book chapters on technical communication, writing across the curriculum, collaborative writing, and film. He also edited a special issue on engineering communication for the *Journal of Language and Learning across the Disciplines*. Dr. Youra has lectured and conducted workshops on writing pedagogy and assessment across the United States and abroad. He is currently investigating issues of authorship in scientific research.

MARK ZACHRY is Associate Professor of Technical Communication at the University of Washington. He has taught a range of courses over the past decade, including theoretical foundations of technical communication, usability testing, user-centered design, and research methods. A focus of his recent research work has been on the development of meaningful models of human-computer interaction in organizations. A related strand of his research develops theoretical constructs to account for the uses of discourse in the regulation of individual practices in organizations. As curriculum chair of the Professional and Technical Writing program at Utah State University, Dr. Zachry worked with his colleagues to explore and refine the uses of assessment techniques to support program development.

Index

For details on these titles from Baywood's Technical Communications Series, visit http://baywood.com.